Annals of Mathematics Studies
Number 24

ANNALS OF MATHEMATICS STUDIES

Edited by Emil Artin and Marston Morse

1. Algebraic Theory of Numbers, *by* HERMANN WEYL
3. Consistency of the Continuum Hypothesis, *by* KURT GÖDEL
6. The Calculi of Lambda-Conversion, *by* ALONZO CHURCH
7. Finite Dimensional Vector Spaces, *by* PAUL R. HALMOS
10. Topics in Topology, *by* SOLOMON LEFSCHETZ
11. Introduction to Nonlinear Mechanics, *by* N. KRYLOFF *and* N. BOGOLIUBOFF
14. Lectures on Differential Equations, *by* SOLOMON LEFSCHETZ
15. Topological Methods in the Theory of Functions of a Complex Variable, *by* MARSTON MORSE
16. Transcendental Numbers, *by* CARL LUDWIG SIEGEL
17. Problème Général de la Stabilité du Mouvement, *by* M. A. LIAPOUNOFF
19. Fourier Transforms, *by* S. BOCHNER *and* K. CHANDRASEKHARAN
20. Contributions to the Theory of Nonlinear Oscillations, Vol. I, *edited by* S. LEFSCHETZ
21. Functional Operators, Vol. I, *by* JOHN VON NEUMANN
22. Functional Operators, Vol. II, *by* JOHN VON NEUMANN
23. Existence Theorems in Partial Differential Equations, *by* DOROTHY L. BERNSTEIN
24. Contributions to the Theory of Games, Vol. I, *edited by* H. W. KUHN *and* A. W. TUCKER
25. Contributions to Fourier Analysis, *by* A. ZYGMUND, W. TRANSUE, M. MORSE, A. P. CALDERON, *and* S. BOCHNER
26. A Theory of Cross-Spaces, *by* ROBERT SCHATTEN
27. Isoperimetric Inequalities in Mathematical Physics, *by* G. POLYA *and* G. SZEGO
28. Contributions to the Theory of Games, Vol. II, *edited by* H. KUHN *and* A. W. TUCKER
29. Contributions to the Theory of Nonlinear Oscillations, Vol. II, *edited by* S. LEFSCHETZ
30. Contributions to the Theory of Riemann Surfaces, *edited by* L. AHLFORS *et al.*

CONTRIBUTIONS TO THE THEORY OF GAMES

VOLUME I

H. F. BOHNENBLUST
G. W. BROWN
M. DRESHER
D. GALE
S. KARLIN
H. W. KUHN
J. C. C. McKINSEY
J. F. NASH
J. VON NEUMANN
L. S. SHAPLEY
S. SHERMAN
R. N. SNOW
A. W. TUCKER
H. WEYL

Princeton, New Jersey
Princeton University Press
1950

London: Geoffrey Cumberlege, Oxford University Press
Printed in the United States of America

First Printing, 1950
Second Printing, 1952

The papers in this volume numbered 1, 4, 7, 8, 9, and 10 were prepared under contracts with the Office of Naval Research.

The papers numbered 3, 5, 6, 11, 12, 13, 14, and 15 were prepared under contracts with the RAND Corporation.

PRINCETON MATHEMATICAL SERIES

Edited by Marston Morse and A. W. Tucker

PREFACE

The foundations of a mathematical theory of "games of strategy" were laid by John von Neumann in several successive stages between 1928 and 1941. The climax of the pioneering period of development came in 1944 with the publication of the THEORY OF GAMES AND ECONOMIC BEHAVIOR by John von Neumann and Oskar Morgenstern. Emphasizing a new approach to competitive economic behavior through a mathematical reduction to suitable games of strategy, this impressive work laid bare a host of problems in the mathematical theory of games. The present study comprises a collection of contributions to this theory and answers some questions raised explicitly or implicitly by von Neumann.

In order to describe the organization and content of this volume, it will be necessary to clarify certain concepts associated with a game which are confused and ambiguous in common use. A *game* is simply the set of rules which describe it, while every particular instance in which the game is played from beginning to end is a *play* of that game. A similar distinction is drawn between the occasion of the selection of one among several alternatives, to be made by one of the players or by some chance device, which is called a *move*, and the actual *choice* made in a particular play. A game consists of a sequence of moves, while a play consists of a sequence of choices.

The decisive step in the mathematical treatment of games is the normalization achieved by the introduction of pure strategies. A *pure strategy* is a plan formulated by a player prior to a play, which will cover all of the possible decisions which may confront him during any play permitted by the rules of the game. Since the pure strategy will provide each choice as a function of the information available to the player at the time of that choice, no freedom of action is lost. Thus the expected course of a play is completely determined by the selection of a pure strategy by each player in ignorance of that chosen by any other player. von Neumann first considered games with a finite number of pure strategies, i.e., *finite* games, and the first part of this volume is devoted to these games. The theory was extended to games with an infinite number of strategies by Jean Ville and Abraham Wald; the contributions of the second part deal with these games, the *infinite* games.

PREFACE

Except for the last two papers of the first section, this study is restricted to zero-sum two-person games. Such games are played by two players and what one player wins, the other loses. By the preceding discussion, a zero-sum two-person game with a finite number of pure strategies can be described by an m by n matrix $A = (a_{ij})$ where each entry a_{ij} represents the amount that player II pays to player I if I uses his ith pure strategy and II uses his jth pure strategy. However, the simplest games, e.g., Matching Pennies, demonstrate that a player is at a disadvantage if he always uses the same pure strategy. Instead, during a long sequence of plays, he may do well to mix his various pure strategies in a random manner but with chosen relative frequencies. This leads – when we pass from such a statistical "sample" of plays of the game to the underlying "population" – to the notion of a mixed strategy as a probability distribution over the pure strategies, with one component for each pure strategy. Mixed strategies for players I and II will be denoted by $x = (x_1, x_2, \ldots, x_m)$ and $y = (y_1, y_2, \ldots, y_n)$, respectively. The set of all mixed strategies for I is an $(m - 1)$-dimensional simplex determined by

$$(1) \qquad x_1 \geq 0,\ x_2 \geq 0,\ \ldots,\ x_m \geq 0 \quad \text{and} \quad x_1 + x_2 + \ldots + x_m = 1 \ ;$$

similarly, the set of all mixed strategies for II is an $(n - 1)$-dimensional simplex determined by

$$(2) \qquad y_1 \geq 0,\ y_2 \geq 0,\ \ldots,\ y_n \geq 0 \quad \text{and} \quad y_1 + y_2 + \ldots + y_n = 1 \ .$$

When player I plays a mixed strategy x and player II a mixed strategy y, the expected payment by II to I is given by the function

$$\phi(x, y) = \sum \sum x_i\, a_{ij} y_j \ .$$

The starting point of all discussion of zero-sum two-person finite games is the Main Theorem [von Neumann-Morgenstern, p. 153] which asserts:

$$\operatorname{Max}_x \operatorname{Min}_y \phi(x, y) = \operatorname{Min}_y \operatorname{Max}_x \phi(x, y) \ .$$

The unique minimax value of ϕ is called the value of the game and will be denoted by v. Mixed strategies x^* and y^* such that $\phi(x^*, y) \geq v \geq \phi(x, y^*)$ for all mixed strategies x and y are called optimal (or good). In these terms the Main Theorem can be given the following form:

Any zero-sum two-person game $A = (a_{ij})$ has a value v and non-empty sets of optimal mixed strategies x^* and y^* characterized by the following inequalities (in addition to (1) and (2), above):

$$(3) \qquad \sum x^*_i a_{i1} \geq v,\ \sum x^*_i a_{i2} \geq v,\ \ldots,\ \sum x^*_i a_{in} \geq v \ ,$$

$$(4) \qquad \sum a_{1j} y^*_j \leq v,\ \sum a_{2j} y^*_j \leq v,\ \ldots,\ \sum a_{mj} y^*_j \leq v \ .$$

PREFACE

The nature of the Main Theorem as a problem in systems of linear inequalities is thus made quite evident. Recognizing this fact and the intimate connection between the problems of linear inequalities and the geometry of convex polyhedra, the first paper in this volume makes available in translation a valuable source paper by Hermann Weyl. This work centers on a proof of the equivalence of definition of the convex closure of a finite set of points in Euclidean space as (1) the set of centers-of-gravity of those points and (2) the intersection of a finite number of half-spaces containing the set of points. The second contribution by Weyl continues in this vein and presents a completely algebraic proof of the Main Theorem, showing that if the entries of the payoff matrix are taken from an ordered field (such as the rational numbers), then the value of the game and the components of the extreme optimal strategies also lie in that field. Alternative algebraic proofs of the Main Theorem are contained in the paper of Shapley and Snow and that of Gale, Kuhn and Tucker on symmetric games.

Considered as subsets of m- and n-dimensional Euclidean spaces, the sets of optimal strategies are easily seen to be closed convex polyhedra. Thus to find all solutions we need only search for the extreme points, termed <u>basic solutions</u> by Shapley and Snow. In their paper they establish an effective procedure for doing this in a finite number of steps. Unfortunately, from a practical point of view, their procedure involves inverting a substantial fraction of the square sub-matrices of the payoff matrix.

It is natural to ask which pairs Π_1, Π_2 of closed convex polyhedra can serve as the sets of optimal strategies for some game. This question is answered completely in this study by Bohnenblust, Karlin and Shapley and also, independently, by Gale and Sherman. Following the paper of Gale and Sherman, we define a pure strategy as <u>essential</u> or <u>superfluous</u> according as it appears with positive probability in some optimal mixed strategy or with zero probability in every optimal mixed strategy. Clearly the numbers e_1, e_2 and s_1, s_2 of essential and superfluous pure strategies for players I and II, respectively, are such that

$$(5) \qquad e_1 > 0,\ e_1 + s_1 = m \text{ and } e_2 > 0,\ e_2 + s_2 = n .$$

If we denote by d_1 the dimension of the optimal-strategy polyhedron Π_1, we can define <u>bounding faces</u> of Π_1 as $(d_1 - 1)$-dimensional faces of Π_1 which do not lie in the boundary of the $(m - 1)$-dimensional simplex of all first-player mixed strategies. We define the bounding faces of the optimal-strategy polyhedron Π_2 similarly and let f_1 and f_2 be the numbers of bounding faces in the two cases. Then Gale and Sherman prove:

CHARACTERIZATION THEOREM. Necessary and sufficient conditions that Π_1 and Π_2 be the sets of optimal strategies for an m by n game are given by

$$e_1 - d_1 = e_2 - d_2 \tag{6}$$

$$f_1 \leq s_2, \quad f_2 \leq s_1 \tag{7}$$

and, of course, by (5) above.

Bohnenblust, Karlin and Shapley prove the same theorem in their paper on the solutions of finite games. In addition, they show that the games with unique solutions ($d_1 = d_2 = 0$) are open and dense in the space of all finite games and present several counterexamples to an analogous characterization theorem for infinite games.

The next group of papers in this study are contributions towards the solution of the principal outstanding problem of zero-sum two-person games: the practical computation of the value and optimal mixed strategies for games with large numbers of pure strategies. In the initial paper of the group, Brown and von Neumann relate a symmetric game (i.e., a game with skewsymmetric payoff matrix) to a system of ordinary differential equations so that optimal strategies arise as limit points of integral curves. This relationship can easily be converted into an iterative procedure for solving games. The existence of such a computational procedure naturally leads to increased interest in methods of symmetrizing an arbitrary game. The paper by Gale, Kuhn and Tucker, which follows, compares two methods of symmetrization and points out that these yield an interesting algebraic proof of the Main Theorem. A second paper by the same authors treats another aspect of the computational problem — the reduction of game matrices to smaller matrices by combining rows or columns (or both) in positive linear combinations, leading to substantial shortcuts in the solutions of many games. In the final work in this group, H. W. Kuhn illustrates many of the new results in this study by describing the solution of a particular game patterned on Poker.

The first part of the volume concludes with two papers on n-person games, the first by Nash and Shapley discussing the solutions of a simple 3-person Poker and that of J. C. C. McKinsey dealing with isomorphism, strategic equivalence, and S-equivalence of zero-sum n-person games. Two games are defined to be <u>isomorphic</u> if there exists a one-to-one correspondence between the imputations of one game and those of the other which preserve the relation of domination [see the THEORY OF GAMES AND ECONOMIC BEHAVIOR for the notions of "imputation" and "domination"]. Two games are <u>strategically equivalent</u> if the relative bargaining positions of the players (and hence their tendencies to form coalitions) are the same for both games. Two games with characteristic functions v and $\overline{v}$ [see loc. cit.] are S-<u>equivalent</u> if there is a positive k, and n numbers $a_1, \ldots, a_n$, whose sum is zero, such that, for every subset S of the set of players

$$\overline{v}(S) = kv(S) + \sum a_i \quad (\text{summed for } i \in S) .$$

Here k changes the monetary unit, and the a_i are fixed payments made at the end of the play regardless of the coalitions formed. McKinsey proves:

$$\text{strategic equivalence} \rightleftarrows \text{isomorphism} \rightleftarrows \text{S-equivalence} .$$

In the zero-sum two-person games discussed in the second part of this study, the two players choose pure strategies u and v from <u>infinite</u> sets U and V. The payoff is the value of a function $k(u, v)$ and mixed strategies are probability distributions μ and ν extended over U and V. Now player I's expectation is given by the double integral

$$\Phi(\mu, \nu) = \iint k(u, v)\, d\mu(u)\, d\nu(v) .$$

The question of the existence of a unique minimax <u>value</u>

$$\text{(8)} \qquad \text{Max}_\mu \text{Min}_\nu \Phi(\mu, \nu) = \text{Min}_\nu \text{Max}_\mu \Phi(\mu, \nu)$$

immediately presents itself for such games. Jean Ville, who was the first to consider this question, proved the general validity of (8) for continuous kernels $k(u, v)$ with U and V unit intervals $0 \leq u \leq 1$, $0 \leq v \leq 1$; subsequent extensions were obtained by Abraham Wald. In the first paper on infinite games in this study Samuel Karlin makes a fresh attack on the value question, using methods from point-set topology, measure theory, and the theory of convex sets. Among many other more detailed results, he proves under very weak assumptions that every game has a value if finitely additive measures are used. Further approaches to the value question are made by Bohnenblust and Karlin in the paper following, which gives generalizations to infinite games of the early proofs by Ville and Kakutani of the Main Theorem for finite games.

The last two papers of the study treat two large categories of infinite games, polynomial games and convex games. In the <u>polynomial</u> games studied by Dresher, Karlin and Shapley, the sets U and V of pure strategies are unit intervals $0 \leq u \leq 1$ and $0 \leq v \leq 1$, and the payoff functions $k(u, v)$ are polynomials in u and v. By applying the geometry of moment spaces to these games, the authors show that the spaces of mixed strategies are finite dimensional and hence that the optimal mixed strategies are composed from a finite number of pure strategies. A bound is given in terms of the degree of the polynomial $k(u, v)$. Results analogous to the Characterization Theorem, above, are obtained for polynomial games using the convex cones found so useful in establishing that theorem. Somewhat weaker theorems hold for polynomial-like games in which

$$k(u, v) = \sum \sum a_{ij} r_i(u) s_j(v) ,$$

where r_i and s_j are continuous functions on the unit interval.

For convex games — i.e., games in which the payoff function $k(u, v)$ is convex in one of the variables — Bohnenblust, Karlin and Shapley find that the convex player must have a pure strategy that is optimal while the other player has optimal mixed strategies composed from a finite number of pure strategies. A second method generalizing a technique for computing solutions of discrete games gives a partially constructive way of obtaining these optimal strategies.

* * *

Covering the rapidly expanding frontier of the theory of games as they do, the contributions of this study provide signposts to present and future trends in research. Some of the outstanding problems which are indicated admit an explicit formulation while many others lie in zones of the theory which need further clarification and restatement above all. Until now, the base of the theory has been the theory of the finite zero-sum two-person game which has received the most intensive development and whose results still influence the direction of research in the other branches. In this portion of the theory, one problem outshadows all others:

(1) To find a computational technique of general applicability for finite zero-sum two-person games with large numbers of pure strategies.

A related question is:

(2) Are there canonical forms for the payoff matrices of such games?

The situation with regard to games with an infinite number of strategies is far less satisfactory. First and foremost, important questions remain to be answered concerning the existence of a value for such games. Among these, we note the following:

(3) To formulate minimax concepts for non-compact sets and unbounded payoff functions. In particular, what are sufficient conditions under which a game with payoff $k(x, y)$ possesses a solution if x and/or y are to be chosen from the set of non-negative real numbers or simply from the set of real numbers?

(4) To supplement the known sufficient conditions for the existence of solutions of infinite games by effective necessary conditions.

Once the existence of a value has been established for a class of infinite games, one naturally desires to extend the structural theorems which hold for finite games.

(5) To find structural theorems for the solutions of comprehensive categories of infinite games, which go beyond the polynomial-like games.

(6) How does one select the best from among the optimal strategies in the case of infinite games?

Naturally, the computation problem occurs with new difficulties in the infinite case.

(7) To find a computational technique of general applicability for zero-sum two-person games with infinite sets of strategies. A constructive method for obtaining the optimal strategies for polynomial-like games or some large class of non-trivial continuous games would constitute a considerable contribution to this problem.

The sparcity of contributions in this study to the theory of the n-person game is an indication not of the completeness of the theory, but rather of wide range of topics still untouched. To the many problems posed by the von Neumann-Morgenstern concept of a solution in terms of coalition we now have added the new problems of the Nash equilibrium point. In this domain we find the following topics to be of interest:

(8) To establish the existence of a solution (in the sense of von Neumann-Morgenstern) of an arbitrary n-person game. It would also be desirable to determine the structural characteristics of individual solutions and of the set of all solutions (of a given game). One suspects that in both cases one deals with sums of a finite number of very simple parametrizing manifolds.

(9) To study n-person games with restrictions imposed on the forming of coalitions (e.g., embargo on side payments, or on communication, or on both), thus recognizing that the cost of communication among the players during the pregame coalition-forming period is not negligible but rather, in the typical economic model with large n, is likely to be the dominating

consideration. One approach to this question might be to formalize the coalition-forming period as a non-cooperative game, in the sense of Nash.

(10) To ascribe a formal "value" to an arbitrary n-person game, even though the operational meaning of the "value" were lacking or obscure. (Cf., the formal value that Karlin defines for arbitrary zero-sum two-person games by means of finitely additive distributions.)

(11) To establish significant asymptotic properties of n-person games, for large n.

It is in the n-person theory that we find the zone of twilight and problems which await clear delineation. Often these problems have their roots in an undeveloped part of the two-person theory. A prime example is the extensive form of a game.

(12) To develop a comprehensive theory of games in extensive form with which to analyze the role of information — i.e., the effect of changes in the pattern of information. At present, equivalence of information patterns can only be defined for games with a value, thus excluding most n-person games.

(13) To develop a dynamic theory of games: (i) In a single play of a multimove game, predict the continuation of the opponent's strategy from his early moves. (ii) In a sequence of plays of the same game, predict the opponent's mixed strategy from his early choices of pure strategies.

The hypothesis of a non-transferable utility has far-reaching effects on the n-person theory. An adequate definition of solution is required for this case.

(14) How does one characterize and find the solutions of games in which each player desires to maximize some non-linear utility-function of the payoff.

If this volume ignores the economic aspects of the theory of games it is not due to a lack of problems but rather to the forced necessity of specialization. A forthcoming Cowles Commission Monograph, "Activity Analysis of Production and Allocation," John Wiley and Sons, is devoted to various facets of a closely related problem of normative economics: the best allocation of limited means toward desired ends.

For the newcomer to the theory of games and the versed researcher as well, a bibliography has been appended to this volume. The novice should find particularly useful a selection of key source papers presenting the type of mathematical background which seems to be applicable to the problems arising in the theory. All readers should find in the wide diversity of topics included there evidence of the vigorous growth of the theory of games as a child of both mathematics and economics.

H. W. Kuhn
A. W. Tucker

Princeton University
and Stanford University
April, 1950

CONTENTS

Part I

FINITE GAMES

THE ELEMENTARY THEORY OF CONVEX POLYHEDRA*

H. Weyl[1]

[A translation by H. W. Kuhn of "Elementare Theorie der konvexen Polyeder," Commentarii Mathematici Helvetici 7 (1935), 290-306.]

§1. THE FUNDAMENTAL THEOREM ON CONVEX PYRAMIDS

If S is a closed bounded point set in (n - 1)-dimensional affine space with the coordinates $x_1, x_2, \ldots, x_{n-1}$, then the points of the convex closure of S can be characterized in two ways: (1) they are the centers of gravity of points from S; (2) they belong to all of the "supports" of S. A support of S is a halfspace

$$\alpha_1 x_1 + \cdots + \alpha_{n-1} x_{n-1} + \alpha \geq 0 ,$$

in which lie all of the points of S. The fundamental theorem concerning convex closures asserts that the two definitions are equivalent. Moreover, (1) can be sharpened so that only centers of gravity of at most n points of S need be admitted, and (2) so that merely the "extreme" supports need be considered. Naturally, the proof of this theorem is obtained with the help of set-theoretic methods; perhaps the simplest exposition is to be found in the introductory part of the paper of Caratheodory, "Uber den Variabilitätsbereich der Fourier'schen Konstanten von positiven harmonischen Funktionen," Rend. Circ. Mat. Palermo 32, 1911, pp. 198-201.

If S consists of a finite number of points only, then the convex closure is a convex polyhedron. It must be possible to derive the fundamental theorem for this case by finite methods; the usual proof does not do this since it entails the application of set-theoretic methods of proof to the convex closure as defined by (1). There appears to be a gap here in the literature which should be filled; it is for this reason that I publish this little note which I had occasion to write down in the summer of 1933 for my last seminar in Göttingen, which had as its subject convex bodies. What we will consider could also be called an elementary theory of finite systems of linear inequalities. It is convenient to proceed from the homogeneous formulation.

A point a in R_n is a sequence of n real numbers $(a_1, a_2, \ldots, a_n)$. Two points a and b, both different from 0, lie on

[1]Received October 24, 1949 by the ANNALS OF MATHEMATICS and accepted for publication; transferred by mutual consent to ANNALS OF MATHEMATICS STUDY No. 24.

*Translation done under a contract with the Office of Naval Research.

the same ray if the b_i are obtained from the a_i through multiplication by a common positive factor of proportionality; such points need not be distinguished in the following. All points x, which satisfy an inequality

(1) $$\alpha_1 x_1 + \dots + \alpha_n x_n \geq 0 ,$$

form a halfspace, which is represented by the "point" $\alpha = (\alpha_1, \dots, \alpha_n) \neq 0$ in the dual space P_n. Again, α is not changed if all of the α_i are multiplied by the same positive factor.

Let a finite system S of points a be given. It is assumed to be non-degenerate; that is, we require that the points a do not all lie in the same hyperplane or, equivalently, do not all satisfy a linear equation

(2) $$\alpha_1 x_1 + \dots + \alpha_n x_n = 0 \quad [(\alpha_1, \dots, \alpha_n) \neq (0, \dots, 0)] .$$

The halfspace determined by (1) is a support for S if all of the points x of the system S satisfy that inequality. It is an extreme support if equality holds for $n - 1$ linearly independent points x of S. There are only a finite number of extreme supports for S; they can be found by choosing all possible combinations of $n - 1$ linearly independent points from S, then testing whether the two halfspaces,

$$\pm (\alpha_1 x_1 + \dots + \alpha_n x_n) \geq 0 ,$$

belonging to the unique hyperplane (2) which passes through each such combination, are supports.

THEOREM 1. (Fundamental theorem.) Let a finite, non-degenerate system of points S be given. A point x, which satisfies all of the extreme support inequalities of S, can be expressed as a positive[2] linear combination of the points $a, b, \dots$ of the system S:

(3) $$x_i = \lambda a_i + \mu b_i + \dots \quad [\lambda \geq 0, \mu \geq 0, \dots] .$$

The totality of points x, which belong to all of the supports, forms a figure which may be called a convex pyramid. A point x, which can be obtained as a positive linear combination (3) of the points $a, b, \dots$ of the system S will be said to be "expressible by S." In non-homogeneous space there always exists at least one support; that is a non-trivial partial result of the fundamental theorem. In the homogeneous space now under discussion, however, it is clearly possible that S have no extreme support at all; then the main theorem says that every point is expressible by S. In the proof, this case must be handled separately.

---

[2]Translator's note: A quantity is called positive if ≥ 0 and strictly positive if > 0.

§ 2. PROOF OF THE FUNDAMENTAL THEOREM

(a) First case: there exist extreme supports.

Let α, β, ... be the extreme supports. Then the inequalities

$$(\alpha e) = \alpha_1 e_1 + \dots + \alpha_n e_n > 0, \quad (\beta e) > 0, \dots$$

hold for

$$e = a + b + \dots ,$$

the "centroid" of S. Let $x = p$ be a point belonging to all of the extreme supports,

$$(\alpha p) \geq 0, \quad (\beta p) \geq 0, \dots .$$

We want to prove that p is expressible. Form $q = p - \lambda e$. The extreme support inequalities are still satisfied for q as long as we have

$$(\alpha p) - \lambda(\alpha e) \geq 0, \quad (\beta p) - \lambda(\beta e) \geq 0, \dots .$$

Therefore, we choose for λ the smallest among the numbers

(4) $$(\alpha p)/(\alpha e), \quad (\beta p)/(\beta e), \dots .$$

Say, for example, $\lambda = (\alpha p)/(\alpha e)$. If q, which lies on an extreme supporting hyperplane: $(\alpha q) = 0$, is expressible, then p is also.

It is good to modify this first stpe in the proof slightly for later purposes. One takes for e not the centroid of S but rather a point appropriately chosen from among a, b, ... Let β be an extreme support. Not all of the points x = a, b, ... of the system S satisfy the equation $(\beta x) = 0$, say $(\beta a) > 0$. Then I choose e = a. The extreme supports fall into two classes: the supports α in the first class satisfying $(\alpha e) > 0$, the supports α_0 in the second class satisfying $(\alpha_0 e) = 0$. The first class is not empty since β, with which we started, belongs to it. Let λ be the minimum of the numbers determined by (4) in which α, β, ... signify all of the supports in the first class. Again we form $q = p - \lambda e$. The essential fact is that q satisfies all of the extreme support inequalities and at least one support equation: $(\alpha q) = 0$.

One can assume that the equation $(\alpha x) = 0$ has the form:

$$x_n = 0 \ (\alpha_1 = \dots = \alpha_{n-1} = 0, \alpha_n = 1) .$$

The point q lies in the supporting hyperplane R_{n-1} with the coordinates $(x_1, \dots, x_{n-1})$ determined by this equation. The proof of the fundamental theorem will be made by proceeding from n - 1 to n, assuming that it already holds for R_{n-1} which we have just introduced. All points x of S satisfy the inequality $x_n \geq 0$. We unite in S_0 those points for which $x_n = 0$ and place the rest in S'. We know that S_0 contains n - 1 linearly independent points since indeed $x_n \geq 0$ was an extreme support. We now argue in the space R_{n-1} for the non-degenerate system S_o. Let

$$(5) \qquad \beta_1 x_1 + \dots + \beta_{n-1} x_{n-1} \geq 0$$

be any extreme support for S_o. I contend that q satisfies this inequality. To see this I form the inequality

$$(6) \qquad \beta_1 x_1 + \dots + \beta_{n-1} x_{n-1} - \mu x_n \geq 0 \ .$$

If it is satisfied for all of the points of S' then it holds for all of the points of S. Therefore I take for μ the minimum of

$$(\beta_1 x_1 + \dots + \beta_{n-1} x_{n-1})/x_n$$

where x runs through the finite number of points in S'; let the minimum be assumed for x = a. Then the inequality (6) is a support for S and indeed an extreme support. For equality holds for n - 2 independent points of S_o and for the point a of S' (which does not lie in R_{n-1}). Therefore q does satisfy the inequality (6) and hence (5). Consequently, by the fundamental theorem, q is expressible by S_o and therefore p is expressible by S. It is inessential to this line of thought whether the (n - 1)-dimensional point system S_o has extreme supports or not.

(b) Second case: there exist no extreme supports.

We start with any halfspace λ:

$$(7) \qquad (\lambda x) \geq 0 \ ,$$

such that its hyperplane $(\lambda x) = 0$ passes through n - 1 linearly independent points of S. By assumption there is at least one point x = e of S on the reverse side: $(\lambda e) < 0$. We will give a construction which replaces the halfspace λ by another containing at least one more point of S. By continuing this process we return to case (a).

Choose the coordinate system so that $(\lambda x) = x_n$ and e has the coordinates (0, 0, ..., 0, -1). Those points of S with their last coordinate $x_n \geq 0$ form a subsystem S^+ of S. Project the points of S^+ from e onto the separating hyperplane $(\lambda x) = 0$; in this manner one obtains a certain system S_o of points in the hyperplane R_{n-1}: $x_n = 0$. Indeed,

$$a = (a_1, \dots, a_{n-1}, a_n) \quad \text{with} \quad a_n \geq 0$$

is projected onto

$$\bar{a} = (a_1, \dots, a_{n-1}) \quad \text{in} \quad R_{n-1} \ .$$

The point $\bar{a}$ is expressible by S, more exactly by (S^+, e): $\bar{a} = a + a_n \cdot e$. If

(8) $$(\alpha x) \equiv \alpha_1 x_1 + \dots + \alpha_{n-1} x_{n-1} \geq 0$$

is an extreme support for S_0, then all of the points of S^+ satisfy this inequality and also e. The hyperplane $(\alpha x) = 0$ passes through $n - 1$ independent points of S since it contains $n - 2$ such points whose projections $\bar{a}$ are independent in R_{n-1} and the point e which does not lie in R_{n-1}. Therefore the halfspace (8) actually contains at least one more point than the halfspace (7) from which we started.

This process fails, however, when S_0 has no extreme supports. But in this case <u>every</u> point in the hyperplane R_{n-1} is expressible by S_0 (since we have assumed the fundamental theorem to be valid in $n - 1$ dimensions) and consequently by (S^+, e). Adding a non-negative multiple of e, we see that we can express all points of the halfspace $(\lambda x) \leq 0$ in this manner. If there is any point e' in S with its coordinate $x_n > 0$, then by adding a positive multiple of e' one obtains an expression for all of the points of the halfspace $(\lambda x) \geq 0$ by (S^+, e, e'). Such an e' must exist since otherwise all of the points of S would satisfy the inequality $x_n \leq 0$ and hence we would have $- x_n = - (\lambda x)$ as an extreme support, contrary to assumption (b).

COROLLARY. We can characterize case (b), in which there exist no extreme supports, by the fact that 0 is expressible: $0 = \lambda a + \mu b + \dots$ with coefficients $\lambda, \mu, \dots$ which are all strictly positive.

Indeed, in case (b) every point is expressible. Adding an expression for $-e$ to the centroid $e = a + b + \dots$, one obtains an expression for 0 with all of the coefficients ≥ 1. The converse is trivial.

THEOREM 2. (Refinement of the fundamental theorem.) A point belonging to all of the extreme supports can be expressed as a positive linear combination of at most n points from S.

For the <u>proof</u> of this refinement one proceeds by choosing for e in case (a) not the centroid, but rather an appropriate point from S itself. Assuming the validity of Theorem 2 for $n - 1$ dimensions, then q can be expressed by at most $n - 1$ points of S lying in R_{n-1}', and hence p by at most n points of S.

<u>In case (b)</u>, the same induction shows us that the points in $(\lambda x) \leq 0$ are expressible by at most n points of S, namely, by $n - 1$ points of S^+ and e. However, one also needs the point e' for the

halfspace $(\lambda x) \geq 0$ so that, at first, only an expression with at most $n + 1$ points results, that is, by $n - 1$ points of S^+, e and e'.

We now consider this point system S' consisting of $n + 1$ points and apply the argument of case (b) to S' instead of S. All of the points of S' except e have their last coordinate $x_n \geq 0$; by projecting them from e onto the hyperplane $x_n = 0$ we obtain the system S_0' containing n points. All points of S' except e' have their last coordinate $x_n \leq 0$. If S_0' has an extreme support then the first part of case (b) yields an extreme support for all of S' immediately and one proves by case (a) that every point which is expressible by the $n + 1$ points of S' is expressible by merely n of them. On the other hand, if S_0' has no extreme supports, then every point with $x_n \leq 0$ can be expressed by n points of S'; a point p with $x_n \geq 0$, however, can be expressed as a positive linear combination of $n - 1$ points b_i of S_0' and e'. All this is mere repetition; but now a new argument enters. Either all of the points b_i belong to S' and then p is expressed by n points of S'. Or only the first $n - 2$ points b_i belong to S', while the last is that positive linear combination of e' and e which lies in $x_n = 0$. But then p is expressible by $(b_1, \ldots, b_{n-2}, e', e)$.

[I have not been able to eliminate this tortuous detour through S' in base (b).]

§ 3. CONSEQUENCES OF THE FUNDAMENTAL THEOREM; SYSTEMS OF LINEAR HOMOGENEOUS INEQUALITIES

Corresponding to the points $a, b, \ldots$ of the system S we have the system S of linear inequalities:

$$\text{(9)} \qquad S: \quad \begin{aligned} (a\xi) &= a_1\xi_1 + \ldots + a_n\xi_n \geq 0 \\ (b\xi) &= b_1\xi_1 + \ldots + b_n\xi_n \geq 0 \\ &\ldots\ \ldots\ \ldots\ \ldots\ \ldots\ \ldots\ \ldots\ . \end{aligned}$$

ξ is a support for the system of points S if and only if ξ satisfies the inequalities S or, as we will say: if ξ belongs to (S). Here (S) stands for the subset of the dual space P_n defined by the inequalities. We call S non-degenerate if there is no ξ except $\xi = 0$ for which equality holds in all of the inequalities S. We divide the fundamental theorem into two parts: first, we contend that every point p which belongs to all of the supports is expressible by S. The condition is clearly not only sufficient but also necessary since a point which is expressible by S obviously belongs to all supports of S. Taking this into account, one can add secondly, that a point necessarily belongs to all of the supports if it belongs to the extreme supports. Thus, there result the following assertions about the system of linear inequalities S.

THEOREM 3. Let the region (S) in the dual space be defined by the finite set of inequalities S, (9). If $(p\xi) \geq 0$ throughout S then the linear form $(p\xi)$ in the variable ξ can be expressed as a positive linear combination of the linear forms $(a\xi)$, $(b\xi)$, ... of the system S.

LEMMA. We call ξ an extreme solution of the system S if equality holds in $n - 1$ linearly independent inequalities. If S is non-degenerate then $(p\xi) \geq 0$ holds for all ξ in (S) if it holds for the extreme ξ.

The fundamental theorem was proved under the assumption that S was non-degenerate. However, the partial statement of Theorem 3 does not depend on this if we operate in the linear subspace R_m of lowest dimension m, which contains all of the points a, b, ... of the system S. On the other hand, the dimension n occurs explicitly in the partial statement of the Lemma; for this reason, the assumption of non-degeneracy is essential here.

[Applying Theorem 3, not to the system S but rather to Σ below, the Lemma yields:

THEOREM 4. If we assume that S is non-degenerate, then every support for S is expressible by the extreme supports.][3]

Establishment of Duality

I. There are only a finite number of extreme solutions ξ for the inequalities S – naturally, considering

$$(\rho\xi_1, \ldots, \rho\xi_n) \quad (\rho > 0)$$

as the same solution as $(\xi_1, \ldots, \xi_n)$. As before, the extreme solutions may be denoted by $\alpha, \beta, \ldots$. The fact that the point x belongs to the extreme supports is expressed by the system of inequalities "dual" to S:

[3]Added by Professor Weyl.

$$(10) \qquad \Sigma: \quad \begin{array}{l} (\alpha x) = \alpha_1 x_1 + \dots + \alpha_n x_n \geq 0 \\ (\beta x) = \beta_1 x_1 + \dots + \beta_n x_n \geq 0 \\ \dots\dots\dots\dots\dots\dots\dots . \end{array}$$

It should be noted that the following inequalities hold

$$\begin{array}{l} (\alpha a) \geq 0, \quad (\alpha b) \geq 0, \quad \dots , \\ (\beta a) \geq 0, \quad (\beta b) \geq 0, \quad \dots , \\ \dots\dots\dots\dots\dots\dots . \end{array}$$

Thus we can state the Lemma in the following manner:

THEOREM 5. If p lies in (Σ) and π lies in (S) then $(p\pi) \geq 0$.

More precisely:

THEOREM 6. The point p belongs to (Σ) if and only if $(p\xi) \geq 0$ for all ξ in (S).

From this follows the "dual" statement:

THEOREM 7. The point π belongs to (S) if and only if $(\pi x) \geq 0$ for all x in (Σ).

For Theorem 5 guarantees that the inequality $(x\pi) \geq 0$ will be satisfied for all π in (S) and x in (Σ). Conversely, if a fixed π satisfies the inequality $(x\pi) \geq 0$ for all x in (Σ) then, in particular, $(a\pi) \geq 0$, $(b\pi) \geq 0$ hold and hence π belongs to (S).

It was not quite correct to characterize Theorems 6 and 7 as dual to each other. For, while $\alpha, \beta, \dots$ are the extreme solutions of the system of inequalities S, the points $a, b, \dots$ need not be the extreme solutions of the system Σ. To restore complete duality we must prove that the extreme solutions of the system Σ are contained among the points $a, b, \dots$. To this end we require:

II. <u>The characterization of the extreme points $\alpha, \beta, \dots$ within</u> (S):

THEOREM 8. The point π is extreme in (S) if and only if the only decomposition of π into summands belonging to (S) is the trivial one, $\pi = \xi' + \xi'' + \dots$, in which $\xi', \xi'', \dots$ lie on the same ray as π:

$$\xi_i' = \rho'\pi_i, \ \xi_i'' = \rho''\pi_i, \ \ldots \ (\rho' \geq 0, \ \rho'' \geq 0, \ \ldots; \ \rho' + \rho'' + \ldots = 1) \ .$$

PROOF. (a) Let π be one of the extreme solutions $\alpha, \beta, \ldots$. Then there are $n - 1$ independent points a, b, ... in S which satisfy the equations:

$$(a\pi) = 0, \quad (b\pi) = 0, \ \ldots \ .$$

However, the individual summands in

$$(a\pi) = (a\xi') + (a\xi'') + \ldots$$

are ≥ 0 and hence it follows from $(a\pi) = 0$ that

$$(a\xi') = 0, \quad (a\xi'') = 0, \ \ldots \ ;$$

in a similar manner

$$(b\xi') = 0, \quad (b\xi'') = 0, \ \ldots \ ,$$
$$\ldots \ \ldots \ \ldots \ \ldots \ \ldots \ \ldots \ \ldots \ .$$

The $n - 1$ linearly independent equations

$$(a\xi') = 0, \quad (b\xi') = 0, \ \ldots$$

have only the single solution π up to a factor of proportionality; hence we have

$$\xi_i' = \rho'\pi_i, \ \xi_i'' = \rho''\pi_i, \ \ldots \ .$$

The inequality $(\pi c) > 0$ holds for at least one point c of the system S. Since $(\xi' c) \geq 0, \ldots,$ it follows that the factors $\rho', \rho'', \ldots$ are non-negative.

(b) If π admits only a trivial decomposition in (S) then π is one of the extreme solutions. For, applying Theorem 3 not to the system of inequalities S but rather to Σ, we see that a representation

$$\pi_i = la_i + mb_i + \ldots; \ l \geq 0, \ m \geq 0, \ \ldots \ ,$$

is possible. In our case, the equations

$$la_i = \rho\pi_i, \ mb_i = \sigma\pi_i, \ \ldots \ (\rho \geq 0, \ \sigma \geq 0, \ \ldots; \ \rho + \sigma + \ldots = 1)$$

must hold by assumption. One of the factors $\rho, \sigma, \ldots$ is different from 0, say ρ, and hence we have, as asserted

$$\pi_i = \frac{1}{\rho} \cdot a_i \ .$$

III. In order to be able to treat the system of inequalities Σ as dual to the system S we must know that the former, as well as the latter, is non-degenerate. For this purpose we introduce the <u>additional assumption</u>

that (S) contain an <u>interior</u> point, that is, a point ξ^0, which satisfies the inequalities

$$(a\xi^0) > 0, \ (b\xi^0) > 0, \ \ldots .$$

THEOREM 9. If S is non-degenerate and (S) contains an interior point then Σ is also non-degenerate.

Namely, let equality hold in all of the inequalities Σ for $x = p$:

(11) $$(\alpha p) = 0, \quad (\beta p) = 0, \ \ldots .$$

Then p as well as -p belongs to (Σ). Hence the two inequalities

$$(p\xi) \geq 0 \quad \text{and} \quad -(p\xi) \geq 0$$

hold and therefore the equation $(p\xi) = 0$ is satisfied for all ξ in (S). In particular, $(p\xi^0) = 0$. But the point p is expressible by S:

$$p_i = \lambda a_i + \mu b_i + \ldots; \ \lambda \geq 0, \ \mu \geq 0, \ \ldots$$

and hence the equation $(p\xi^0) = 0$ provides us with:

$$\lambda(a\xi^0) + \mu(b\xi^0) + \ldots = 0 .$$

Since the individual factors $(a\xi^0)$, $(b\xi^0)$, ... are positive by assumption, all of the non-negative coefficients λ, μ, ... must vanish. This implies $p = 0$ and hence equations (11) have no solution other than $p = 0$.

We remark that, with the assumption of Theorem 9, which will be retained for the rest of the paper, (Σ) also contains interior points: for example, the centroid $a + b + \ldots$ of S is such an interior point.

IV. Let the extreme solutions of Σ be a', b', ...

THEOREM 10. The region (S') defined by the inequalities

$$S': \ (a'\xi) \geq 0, \ (b'\xi) \geq 0, \ \ldots ,$$

dual to the system Σ, is the same as (S). The points a', b', ... are a subset of S.

The first part of the statement: (S) = (S') follows if one applies Theorem 6 to Σ instead of S and compares with Theorem 7. Since a' belongs to (Σ) we have the expression

$$a'_i = \lambda a_i + \mu b_i + \ldots; \ \lambda \geq 0, \ \mu \geq 0, \ \ldots .$$

However, since a' is extreme in (Σ), it follows (with the help of characterization II above as in the proof of part (b) of II) that a' must be identical with one of the points a, b, ... (up to a factor of proportionality).

Σ was formed from the extreme solutions $\alpha, \beta, \ldots$ of S and, conversely, S' by means of the extreme solutions a', b', ... of Σ. Now we must consider ıe extreme solutions of S': $\alpha', \beta', \ldots$. However, these are not merely a subset of $\alpha, \beta, \ldots$ but are exactly the extreme solutions of S:

THEOREM 11. The systems of inequalities S' and Σ are mutually dual to one another.

For in II the extreme ξ are characterized on the basis of the region (S) of all ξ; however, the regions (S) and (S') are identical. This result may be expressed in the following manner in terms of convex pyramids:

THEOREM 12. Through each extreme edge there pass n - 1 independent supporting hyperplanes, while n - 1 independent extreme edges lie in each extreme supporting hyperplane.

Consequently, it is equivalent to proceed to define a convex pyramid by a finite number of points a, b, ..., as above, or by a finite number of supports $\alpha, \beta, \ldots$.

V. As a consequence of the complete dualization we have:

THEOREM 13. The intersection of two convex pyramids is again a convex pyramid.

To prove this one defines each of the given pyramids by a finite number of support inequalities and then combines both systems of support inequalities into a single one; this system again defines a convex pyramid. If we wish to derive it from a finite number of points a, b, ... then we must choose a, b, ... as the extreme solutions of the combined system of support inequalities.

All of these results are trivial after the proof of the fundamental theorem. The service that our exhaustive treatment should render is merely the enumeration of these consequences in their proper order, in which they follow logically from each other.

§4. CONVEX POLYHEDRA, INHOMOGENEOUS LINEAR INEQUALITIES

I. <u>A convex polyhedron as the convex closure of a finite system of points.</u>

On setting $x_n = -1$ in the homogeneous space R_n there results an inhomogeneous $(n - 1)$-dimensional space $\bar{R}_{n-1}$. If S is a finite, non-degenerate system of points $a, b, \ldots$ in $\bar{R}_{n-1}$ $(a_n = b_n = \ldots = -1)$, then case (b) of the fundamental theorem, in which every point x is expressible, is impossible. Indeed $-x_n \geq 0$ holds for every expressible point x since we have:

$$x_i = \lambda a_i + \mu b_i + \ldots \quad (\lambda \geq 0, \mu \geq 0, i = 1, \ldots, n) . \tag{12}$$

Hence, there always exists an extreme support. If one wishes the expressions for the points x normalized so that $x_n = -1$, then the non-negative parameters $\lambda, \mu, \ldots$ in (12) must be restricted by the condition $\lambda + \mu + \ldots = 1$. The points in $\bar{R}_{n-1}$ which are expressible by S form the convex closure H of S, the "convex polyhedron" generated by S. It can be defined by the finite system of support inequalities.

Here the additional hypothesis (see III in §3) that the dual system is non-degenerate is fulfilled by Theorem 9, since all of the points $x = a, b, \ldots$ of S satisfy the inequality

$$0 \cdot x_1 + \ldots + 0 \cdot x_{n-1} - 1 \cdot x_n > 0 .$$

If $\alpha_1, \ldots, \alpha_{n-1}$ are numbers given arbitrarily, form

$$\min(\alpha_1 x_1 + \ldots + \alpha_{n-1} x_{n-1}) = \alpha_n ,$$

where $x = (x_1, \ldots, x_{n-1})$ runs through the finite number of points in S; then

$$\alpha_1 x_1 + \ldots + \alpha_{n-1} x_{n-1} - \alpha_n \geq 0 \tag{13}$$

is a support for S. Hence, there is a support for S for $\alpha_1, \ldots, \alpha_{n-1}$ prescribed arbitrarily and, indeed, such that its hyperplane passes through a point of S.

We never have $(\alpha_1, \ldots, \alpha_{n-1}) = (0, \ldots, 0)$ for an extreme support (13) of S since its hyperplane contains at least one point of S, so that the vanishing of $\alpha_1, \ldots, \alpha_{n-1}$ would imply $\alpha_n = 0$ also.

If we call the extreme supports, faces, and the extreme solutions of the dual system Σ, edges of H, then the following theorem holds:

> THEOREM. At least $n - 1$ independent faces pass through every edge of the convex polyhedron H and at least $n - 1$ independent edges lie in every face.

II. A convex polyhedron as the intersection of a finite number of halfspaces.

The finite system of inequalities

$$(\alpha x) \equiv \alpha_1 x_1 + \dots + \alpha_{n-1} x_{n-1} - \alpha_n \geq 0, \quad (\beta x) \geq 0, \quad \dots$$

defines a subset H of the space $\overline{R}_{n-1}$. If π: $(\pi x) \geq 0$ is a support for H then, according to the fundamental theorem, π must be expressible in terms of the points $\alpha, \beta, \dots$ of the system Σ. By I, H can only be a convex polyhedron if <u>every</u> point $\pi' = (\pi_1, \dots, \pi_{n-1})$ in the homogeneous space R_{n-1} is expressible by means of the finite set of points

(14) $$\alpha' = (\alpha_1, \dots, \alpha_{n-1}), \quad \beta' = (\beta_1, \dots, \beta_{n-1}), \quad \dots .$$

Hence the system of points (14), Σ', possesses no extreme supports in R_{n-1} (case (b) of the fundamental theorem). In addition, H must have an interior point, that is, there must be a point c in $\overline{R}_{n-1}$ such that

(15) $$(\alpha c) > 0, \quad (\beta c) > 0, \quad \dots$$

hold. But this is also sufficient. To prove this we take the point c as the origin; then we have

$$\alpha_n < 0, \quad \beta_n < 0, \quad \dots .$$

It follows immediately that Σ is non-degenerate, that is, that there are no numbers $(d_1, \dots, d_{n-1}, d_n) \neq (0, \dots, 0, 0)$ for which all of the equations

$$\alpha_1 d_1 + \dots + \alpha_n d_n = 0, \quad \beta_1 d_1 + \dots + \beta_n d_n = 0, \quad \dots$$

subsist. Indeed, otherwise one could derive an equation: $\pi_n d_n = 0$ with a <u>negative</u> coefficient π_n by expressing 0 by the points (14) in R_{n-1} and taking account of the Corollary to the fundamental theorem. Thus, having shown that $d_n = 0$, this contradicts the fact that all points $(\pi_1, \dots, \pi_{n-1})$ are expressible by Σ', not only those which satisfy the equation

$$\pi_1 d_1 + \dots + \pi_{n-1} d_{n-1} = 0 .$$

That the system S dual to Σ is non-degenerate follows from Theorem 9 in consequence of the hypothesis (15). We must still show that S consists of points in $\overline{R}_{n-1}$.

Necessarily, $x_n \leq 0$ for a solution x of the inequalities

(16) $$\alpha_1 x_1 + \dots + \alpha_n x_n \geq 0, \quad \beta_1 x_1 + \dots + \beta_n x_n \geq 0, \quad \dots$$

since the same proof that was applied to the corresponding equations above yields $\pi_n x_n \geq 0$. If the solution x is extreme then $x_n < 0$; otherwise we would have an extreme solution of the inequalities

$$\alpha_1 x_1 + \dots + \alpha_{n-1} x_{n-1} \geq 0, \quad \beta_1 x_1 + \dots + \beta_{n-1} x_{n-1} \geq 0, \quad \dots$$

contrary to hypothesis. Hence one can choose $x_n = -1$ for an extreme solution x so that the homogeneous inequalities (16) reduce to the inhomogeneous inequalities (13): their extreme solutions form a finite system of points S in the inhomogeneous space $\bar{R}_{n-1}$. Moreover, every point x, which satisfies all of the inequalities (13) can be expressed by S; in other words, H is identical with the convex closure of S. The points of S are the vertices of this convex polyhedron.

III. Normal cones.

A point $(\alpha_1, \ldots, \alpha_{n-1})$ in homogeneous R_{n-1} is called a normal to the vertex a of a given polyhedron if

$$\alpha_1 a_1 + \ldots + \alpha_{n-1} a_{n-1} = \min(\alpha_1 x_1 + \ldots + \alpha_{n-1} x_{n-1})$$

or

$$\alpha_1 (x_1 - a_1) + \ldots + \alpha_{n-1}(x_{n-1} - a_{n-1}) \geq 0 \; ; \tag{17}$$

here x runs through all of the vertices $a, b, \ldots$ of the polyhedron. The extreme supporting hyperplanes of the polyhedron which pass through the vertex a provide exactly the extreme solutions of the finite system of inequalities (17). Since there are $n - 1$ linearly independent hyperplanes with this property, the extreme solutions of (17) form a non-degenerate system of points in R_{n-1}. Every normal can be expressed as a positive linear combination of these: the "normal cone" is a non-degenerate convex pyramid in R_{n-1}.

Every point $(\alpha_1, \ldots, \alpha_{n-1})$ belongs to the normal cone of at least one vertex since when x runs through the vertices $a, b, \ldots$ the minimum of $\alpha_1 x_1 + \ldots + \alpha_{n-1} x_{n-1}$ is assumed for at least one of these points. However, the normal cones of different vertices have different interior points. Indeed $(\alpha_1, \ldots, \alpha_{n-1})$ is an interior point of the normal cone of a if proper inequality holds in (17) for $x = b, c, \ldots$; for example,

$$\alpha_1 b_1 + \ldots + \alpha_{n-1} b_{n-1} > \alpha_1 a_1 + \ldots + \alpha_{n-1} a_{n-1} \; .$$

But exactly the opposite inequality,

$$\alpha_1 b_1 + \ldots + \alpha_{n-1} b_{n-1} \leq \alpha_1 a_1 + \ldots + \alpha_{n-1} a_{n-1} \; ,$$

holds for a point belonging to the normal cone of b.

IV. Families of polyhedra.

Following Minkowski, one can form a linear combination, $\lambda_1 H_1 + \ldots + \lambda_h H_h = H$, of several convex polyhedra with positive numerical coefficients λ; H is again a convex polyhedron. The process can be carried out in two steps: (1) Multiplication by a positive factor λ; (2) Addition. As a third step, one may consider the coefficients λ_i as variables in the domain $\lambda_1 > 0$ with a fixed basis $H_1, \ldots, H_h$: then H runs through a <u>family of polyhedra</u>.

<u>Multiplication of the convex polyhedron H by a positive factor</u> λ. Let H be the convex closure of the finite system S of points a, b, c, ... (one can eliminate those points which are not vertices without changing H). Following Minkowski, the minimum which has already been used above

$$\min_{x=a,b,c,\ldots} (\alpha_1 x_1 + \cdots + \alpha_{n-1}x_{n-1}) = h(\alpha_1, \ldots, \alpha_{n-1})$$

will be called the <u>support function</u>. The halfspace

$$\alpha_1 x_1 + \cdots + \alpha_{n-1}x_{n-1} - \alpha_n \geq 0$$

is a support for S (or H) if and only if

$$\alpha_n \leq h(\alpha_1, \ldots, \alpha_{n-1}) .$$

We obtain λH from H by replacing every point $x = (x_1, \ldots, x_{n-1})$ of H by $\lambda x = (\lambda x_1, \ldots, \lambda x_{n-1})$. Then λH is the convex closure of the points λa, λb, λc, ... Apposed to this we have the dual statement: λH is the polyhedron with support function λh. The normal cones of the various vertices λa, λb, λc, ... of λH are the same as the normal cones of the corresponding vertices a, b, c, ... of H.

<u>Addition</u>. Let H be the convex closure of S: a, b, c, ... with the support function h and let H' be the convex closure of S': a', b', c', ... with the support function h'. We define H + H' to contain all points of the form x + x', where x is an arbitrary point of H and x' is an arbitrary point of H'. This transfinite construction can be replaced immediately by the following finite one: the point a + a' is formed of every combination (a, a') of a point a of S and a point a' of S'; in this manner, a system of points S + S' is obtained. The polyhedron H + H' is the convex closure of S + S'. In fact, every positive linear combination

$$\Sigma \mu(a + a') = \Sigma \mu a + \Sigma \mu a' \quad (\mu \geq 0, \ \Sigma \mu = 1)$$

is readily seen to be the sum of a point x of H and a point x' of H'. Conversely, if x belongs to H: $x = \Sigma \mu a$, and x' to H': $x' = \Sigma \mu' a'$ $(\mu \geq 0, \mu' \geq 0; \Sigma \mu = 1 \ \Sigma \mu' = 1)$ then

$$x + x' = \Sigma \mu\mu'(a + a') .$$

Again we have the dual interpretation: H + H' is the convex polyhedron with support function h + h'. In fact, the relation

$$h(\alpha) + h'(\alpha) = \min_{\substack{x = a, b, c, \dots \\ x' = a', b', c', \dots}} \{\alpha_1(x_1 + x_1') + \dots + \alpha_{n-1}(x_{n-1} + x_{n-1}')\}$$

follows from

$$h(\alpha) = \min_{x=a,b,c,\dots} (\alpha_1 x_1 + \dots + \alpha_{n-1}x_{n-1}) ,$$

$$h'(\alpha) = \min_{x'=a';b',c',\dots} (\alpha_1 x'_1 + \dots + \alpha_{n-1}x_{n-1}') .$$

The general theory tells us how to select the finite number of extreme support inequalities which serve to define $H + H'$.

However, by considering the polar figure we learn even more. The homogeneous space R_{n-1} is divided on the one hand into the normal cones of H and on the other, into those of H'. On the basis of Theorem 13, the superposition of these two divisions yields a new division of R_{n-1} into convex pyramids: this is the normal figure of $H + H'$. Indeed, the combination of a vertex a of H with vertex a' of H' only gives rise to a vertex $a + a'$ of $H + H'$ if the normal cones of a and a' have interior points in common; the intersection is the normal cone of $a + a'$ in $H + H'$. The extreme points of the normal cone of $H + H'$ provide the "normals" $(\alpha_1, \dots, \alpha_{n-1})$ to the extreme supports for $H + H'$.

On the basis of these remarks, consider the relations in a "family" of convex polyhedra such as $\lambda H + \lambda' H'$, allowing λ and λ' to vary freely in the domain $\lambda > 0$, $\lambda' > 0$. The normal figure of H is not changed by multiplication by λ. As a consequence, the normal figure of a polyhedron in the family does not depend on the values of λ and λ'; in particular, neither do the normals to the extreme supports nor do those combinations (a, a') of a vertex a of H with a vertex a' of H' which give rise to a vertex $\lambda a + \lambda' a'$ of $\lambda H + \lambda' H'$. Likewise, the combinatorial scheme of vertices and extreme supporting hyperplanes, which tells how the one is distributed upon the other, is constant within the family.

H. Weyl

The Institute for Advanced Study

ELEMENTARY PROOF OF A MINIMAX THEOREM DUE TO VON NEUMANN *

Hermann Weyl

J. von Neumann's minimax problem in the theory of games belongs to the theory of linear inequalities and can be approached in the same elementary way in which I proved the fundamental facts about convex pyramids. As elementary are considered such operations in an ordered field K of numbers as require nothing but addition, subtraction, multiplication and division, and the decision whether a given number is > 0 or $= 0$ or < 0. Decisions about a set of numbers are elementary only if they concern a finite set, the members of which are exhibited one by one. In such a sequence of numbers $\alpha_1, \ldots, \alpha_n$ we can find the smallest, $\min \alpha_k$, and the biggest, $\max \alpha_k$. As for the field K no continuity axioms, not even the axiom of Archimedes, are assumed.

Let there be given a matrix of numbers

$$a_{ik}(i = 1, \ldots, m; k = 1, \ldots, n) .$$

For any point $\eta = (\eta_1, \ldots, \eta_n)$ in n-space set

$$m(\eta) = \min_i (\Sigma_k a_{ik}\eta_k) ,$$

and for any point $\xi = (\xi_1, \ldots, \xi_m)$ in m-space:

$$M(\xi) = \max_k (\Sigma_i a_{ik}\xi_i) .$$

After suitably constructing a number λ_0, a point in n-space $\eta = \eta^0$ satisfying the conditions

$$\eta^o_k \geq 0 (k = 1, \ldots, n), \Sigma_k \eta^o_k = 1$$

and a point in m-space $\xi = \xi^0$ satisfying the corresponding conditions

$$\xi^o_i \geq 0 (i = 1, \ldots, m), \Sigma_i \xi^o_i = 1 ,$$

we are going to establish the following

Fundamental facts:

$$m(\eta) \leq \lambda_0$$

---

*Accepted as a direct contribution to ANNALS OF MATHEMATICS STUDY No. 24.

whenever

(1) $$\eta_k \geq 0 (k = 1, \ldots, n), \Sigma_k \eta_k = 1 ,$$

the upper bound λ_0 being obtained for $\eta = \eta^0$, and

$$M(\xi) \geq \lambda_0$$

whenever

(2) $$\xi_i \geq 0 (i = 1, \ldots, m), \Sigma_i \xi_i = 1 ,$$

the lower bound being obtained for $\xi = \xi^0$.

One part of the main theorem about convex pyramids which I proved in elementary fashion (and independently of the rest) and which I am now going to repeat as Lemma 1 is concerned with a configuration Σ of points

$$b_j = (b_{j1}, \ldots, b_{jn}) \quad (j = 1, \ldots, r)$$

which span n-space, and the extreme supports of this configuration. Given a linear form $(\gamma x) = \gamma_1 x_1 + \ldots + \gamma_n x_n$, let us say that the point x lies in the plane γ or in the half-space γ if $(\gamma x) = 0$ or $(\gamma x) \geq 0$ respectively. A form $\gamma \neq 0$ such that all points b_j of the set Σ lie in the half-space γ is called a support of the set, and this support is extreme if $n - 1$ linearly independent b_j, for instance

$$b_{j_1}, \ldots, b_{j_{n-1}} \quad (j_1 < j_2 < \ldots < j_{n-1}) ,$$

lie in the plane γ. In this case (γx) is proportional to

$$|b_{j_1}, \ldots, b_{j_{n-1}}, x| .$$

LEMMA 1. Either the configuration Σ has an extreme support, or every point p is representable in the form

$$p = \sum_{j=1}^{r} \mu_j b_j \quad (\mu_j \geq 0) .$$

One may cancel here the adjective extreme and then add the corollary: If Σ has a support it has an extreme support.

We are interested in the case where the configuration Σ consists of the n unit points

$$e_1 = (1, 0, \ldots, 0), e_2 = (0, 1, \ldots, 0), \ldots, e_n = (0, 0, \ldots, 1)$$

and some further points

$$a_i = (a_{i1}, \ldots, a_{in}) \quad (i = 1, \ldots, m):$$

$$b_k = e_k (k = 1, \ldots, n); \; b_{n+i} = a_i (i = 1, \ldots, m) .$$

Set $e = e_1 + \ldots + e_n = (1, \ldots, 1)$. Every support γ ($\neq 0$) satisfies the inequalities

$$(\gamma e_k) = \gamma_k \geq 0 \quad (k = 1, \ldots, n)$$

and hence we may normalize by

$$(\gamma e) = \gamma_1 + \ldots + \gamma_n = 1 .$$

If the second alternative of Lemma 1 takes place, represent $-e$ and add

$$e = 1 \cdot e_1 + \ldots + 1 \cdot e_n + 0 \cdot a_1 + \ldots + 0 \cdot a_m .$$

One thus finds a representation of zero,

$$0 = \sum \rho_k e_k + \sum \xi_i a_i \quad \text{with } \rho_k \geq 1, \; \xi_i \geq 0 ,$$

hence

$$\sum_i \xi_i a_{ik} = - \rho_k < 0 \quad (k = 1, \ldots, n) .$$

Thus we may state the following two facts:

LEMMA 2. If the system Σ consisting of the points $e_k (k = 1, \ldots, n)$ and $a_i (i = 1, \ldots, m)$ has a support then it has an extreme support.

LEMMA 3. If it has no support, i.e. if the inequalities

$$\eta_k \geq 0 \quad (k = 1, \ldots, n), \sum_k a_{ik} \eta_k \geq 0 \quad (i = 1, \ldots, m)$$

have none but the trivial solution $\eta = 0$, then the n inequalities

$$\sum_i \xi_i a_{ik} < 0 \quad (k = 1, \ldots, n)$$

are solvable by non-negative ξ_i.

We find it convenient to apply Lemma 3 to the matrix $- ||a_{ki}||$ rather than to $||a_{ik}||$ and to change the notation i, m, ξ into k, n, η. Then Lemma 3 takes on the (nearly identical) form:

LEMMA 4. Either the system

$$\sum_k a_{ik} \eta_k > 0 \quad (i = 1, \ldots, m), \quad \eta_k \geq 0 \quad (k = 1, \ldots, n)$$

or the system

$$\sum_i \xi_i a_{ik} \leq 0 \; (k = 1, \ldots, n), \; \xi_i \geq 0 \; (i = 1, \ldots, m), \; \sum \xi_i = 1$$

is solvable.

One needs but form $\sum_{i,k}\xi_i a_{ik}\eta_k$ in order to see that this is a disjunctive either-or.

Let us now introduce a parameter λ called time and try to find out for which values of λ the system Σ_λ consisting of the n points e_k and the m points $a_i - \lambda e$ has or has not a support. In the first case we say that λ makes good. If λ makes good so does every $\lambda' \leq \lambda$, and the same support γ will do for all of them. Indeed a support γ of Σ_λ satisfies the inequalities

$$\gamma_1 \geq 0, \ldots, \gamma_n \geq 0, (\gamma e) > 0$$

and thus

$$(\gamma, a_i - \lambda' e) = (\gamma, a_i - \lambda e) + (\lambda - \lambda')(\gamma e) \geq 0$$

holds as a consequence of $(\gamma, a_i - \lambda e) \geq 0$ and $\lambda - \lambda' \geq 0$.

Take any point η satisfying (1). For

$$\lambda = \min_i (a_i \eta) = \min_i (\Sigma_k a_{ik}\eta_k) = m(\eta)$$

one has

$$(\eta, a_i - \lambda e) \geq 0 (i = 1, \ldots, m), (\eta e_k) \geq 0 (k = 1, \ldots, n) ;$$

hence this λ makes good. On the other hand, given λ and a $\gamma \neq 0$ satisfying the relations

$$(\gamma, a_i - \lambda e) \geq 0 (i = 1, \ldots, m), \gamma_k = (\gamma e_k) \geq 0 (k = 1, \ldots, n)$$

one finds for every point of the form

$$p = \Sigma_i \xi_i a_i \quad (\xi_i \geq 0, \xi_1 + \ldots + \xi_m = 1)$$

the inequality

$$(\gamma, p - \lambda e) \geq 0 \quad \text{or} \quad \Sigma_k \gamma_k (p_k - \lambda) \geq 0 .$$

Hence at least <u>one</u> of the $p_k - \lambda \geq 0$, or

$$\lambda \leq \max_k p_k = \max_k (\Sigma_i \xi_i a_{ik}) = M(\xi) .$$

Any $M(\xi)$ corresponding to a point ξ fulfilling the relations (2) is therefore an upper bound for those λ that make good.

Use the unified notation

$$\delta_k = 0 \text{ for } k = 1, \ldots, n; \quad \delta_{n+i} = 1 \text{ for } i = 1, \ldots, m;$$
$$b_j(\lambda) = b_j - \lambda\delta_j e \quad (j = 1, \ldots, r = n + m) .$$

In view of Lemma 2 pick any $(n - 1)$-combination $J = \{j_1 < j_2 < \ldots < j_{n-1}\}$ out of the indices $j = 1, \ldots, r$. If $b_{j_1}(\lambda), \ldots, b_{j_{n-1}}(\lambda)$ are linearly independent and the plane going through them is a supporting plane for the entire set Σ_λ of the $b_j(\lambda)$ then we shall say that the combination J is alive at the moment λ. Let us express that the combination[1] $J = \{1, 2, \ldots, n-1\}$ is alive at the moment λ', or that we have a $\gamma \neq 0$ such that

$$(\gamma, b_j(\lambda')) \geq 0 \quad \text{for all } j$$
$$\text{and } = 0 \quad \text{for } j = 1, \ldots, n-1 .$$

This implies $\gamma_k \geq 0$, hence $(\gamma e) > 0$, and we may normalize by $(\gamma e) = 1$. Therefore e is independent of the $n - 1$ linearly independent points $b_1(\lambda'), \ldots, b_{n-1}(\lambda')$, or

$$D = |b_1, \ldots, b_{n-1}, e| = |b_1(\lambda'), \ldots, b_{n-1}(\lambda'), e| \neq 0 .$$

The normalized expression of (γx) is

$$(\gamma x) = \frac{1}{D} |b_1(\lambda'), \ldots, b_{n-1}(\lambda'), x| .$$

Consequently one has to form the r functions

$$\varphi_j(\lambda) = \frac{1}{D} |b_1(\lambda), \ldots, b_{n-1}(\lambda), b_j(\lambda)| .$$

[$\varphi_1(\lambda), \ldots, \varphi_{n-1}(\lambda)$ vanish identically.] If J is alive at the moment λ' the simultaneous inequalities

$$\varphi_j(\lambda') \geq 0 \quad (j = 1, \ldots, r)$$

hold (and vice versa). Notice that

$$|b_1(\lambda), \ldots, b_{n-1}(\lambda), b_n(\lambda)| = \begin{vmatrix} b_{11}, & \ldots, & b_{1n}, & \lambda\delta_1 \\ \ldots & \ldots & \ldots & \ldots \\ b_{n1}, & \ldots, & b_{nn}, & \lambda\delta_n \\ 1, & \ldots, & 1, & 1 \end{vmatrix}$$

is a linear function of λ. Write therefore

$$\varphi_j(\lambda) = \alpha_j^0 - \alpha_j \lambda$$

and distinguish between indices j of the first and second class by $\alpha_j \leq 0$ and $\alpha_j > 0$ respectively.

For an index j of the first class

I) $$\varphi_j(\lambda) \geq \varphi_j(\lambda') \geq 0 \quad \text{for} \quad \lambda \geq \lambda' .$$

---

[1]This combination serves merely as an example for an arbitrary J. Hence the reader should ignore the special value 0 which δ_k takes on for $k = 1, \ldots, n$.

For an index j of the second class the inequality $\varphi_j(\lambda') \geq 0$ states that

$$\lambda' \leq \alpha_j^0/\alpha_j .$$

The second class cannot be vacuous; for then J would stay alive at all times $\lambda \geq \lambda'$, and every λ would make good, which we know to be impossible. Determine the minimum λ_J of the quotients α_j^0/α_j that correspond to the indices j of the second class. Then

II) $$\lambda' \leq \lambda_J$$

and

III) $$\varphi_j(\lambda_J) \geq 0, \; \varphi_j(\lambda) \geq 0 \quad \text{for} \quad \lambda \leq \lambda_J$$

and every j of the second class. The relations I), II), III) show: if J is alive at the moment λ' then $\lambda' \leq \lambda_J$ and J keeps alive during the whole period $\lambda' \leq \lambda \leq \lambda_J$; in other words, the uninterrupted life of J terminates at $\lambda = \lambda_J$.

Rearranging the argument, I describe once more how the "admissible" combinations J are separated from the others which never come to life. The combination $J = \{1, \ldots, n - 1\}$ is ruled out (1) if the determinant D vanishes. Otherwise form the functions $\varphi_j(\lambda) = \alpha_j^0 - \alpha_j \lambda$ and rule out J if (2) all $\alpha_j \leq 0$. But if some α_i are positive take the least of the quotients α_j^0/α_j corresponding to those j for which $\alpha_j > 0$ (j's of the second class) and call it λ_J. The combination J is thrown away unless (3) $\varphi_j(\lambda_J) \geq 0$ for all j, even for those of the first class. A J which passes the three consecutive tests is admissible. Then we may state our result as follows:

> LEMMA 5. If λ makes good then there is an admissible combination J whose death occurs at a time $\lambda_J \geq \lambda$.

Indeed if the system Σ_λ has a support it also has an extreme support determined by $n - 1$ linearly independent points $b_{j_1}(\lambda), \ldots, b_{j_{n-1}}(\lambda)$. The combination $J = \{j_1, \ldots, j_{n-1}\}$ is then alive at the moment λ.

Lemma 5 shows that there are admissible combinations J. Choose now for λ_0 the largest of the λ_J that correspond to the admissible combinations J and let $J = J_0$ be that admissible combination for which $\lambda_{J_0} = \lambda_0$. This $J_0 = \{j_1, \ldots, j_{n-1}\}$ yields an extreme support

$$(\eta^0 x) = \frac{|b_{j_1}(\lambda_0), \ldots, b_{j_{n-1}}(\lambda_0), x|}{|b_{j_1}(\lambda_0), \ldots, b_{j_{n-1}}(\lambda_0), x|}$$

for Σ_{λ_0} (and hence a support η^0 for Σ_λ, $\lambda \leq \lambda_0$). Thus λ_0 makes good, but according to Lemma 5 no $\lambda > \lambda_0$ does.

We have seen that $m(\eta)$ is a λ that makes good whenever the point η satisfies the conditions (1), and therefore $m(\eta) \leq \lambda_0$. The upper bound is actually reached for $\eta = \eta^0$; indeed

$$(\eta^0, a_i - \lambda_0 e) \geq 0 \quad \text{or} \quad \lambda_0 \leq (\eta^0 a_i) \quad (i = 1, \ldots, m), \lambda_0 \leq m(\eta^0) \ .$$

This is the first of our Fundamental facts.

On the other hand there can be no γ such that

$$(\gamma, a_i - \lambda_0 e) > 0 \ (i = 1, \ldots, m), \ (\gamma e_k) \geq 0 \ (k = 1, \ldots, n), \ (\gamma e) = 1 \ .$$

For then $\lambda_1 = \min_i (\gamma a_i) > \lambda_0$ would make good, against our better knowledge. Hence by applying Lemma 4 to the matrix $a_{ik} - \lambda_0$ we find a solution $\xi = \xi^0$ of

$$\Sigma_i \xi_i (a_{ik} - \lambda_0) \leq 0, \ \xi_i \geq 0, \Sigma \xi_i = 1 \ ,$$

so that we have

$$\text{(3)} \qquad M(\xi^0) \leq \lambda_0, \ \xi_i^0 \geq 0 \quad (i = 1, \ldots, m), \ \Sigma \xi_i^0 = 1 \ .$$

But for every point ξ satisfying (2) and every λ that makes good, $\lambda \leq M(\xi)$, in particular $\lambda_0 \leq M(\xi)$. According to (3) the lower bound λ_0 is actually reached for $\xi = \xi^0$. This completes the proof.

H. Weyl

The Institute for Advanced Study

BASIC SOLUTIONS OF DISCRETE GAMES

L. S. Shapley and R. N. Snow[1]

§1. SUMMARY

All solutions of the general two-person zero-sum game can be represented by means of a finite number of "basic solutions," which may be visualized as pairs of vertices from two convex sets in the spaces of all mixed strategies for the two players. Each basic solution has associated with it one or more sub-matrices of the whole game-matrix, called the kernels of the solution. Once a basic kernel has been located in the game-matrix, the associated basic solution may be computed by a formula.

§2. INTRODUCTORY

The game Γ is represented by the $m \times n$ matrix $M = ||a^i_j||$ where a^i_j is the payment to player I by player II, corresponding to their play of strategies i and j respectively, $i = 1, 2, \ldots, m$, $j = 1, 2, \ldots n$. Superscripts will be used to denote rows and subscripts to denote columns of the matrix M, i.e. $\vec{a}^i$ is the ith row, $\vec{a}_j$ the jth column of M. Boldface capitals will denote matrices, while boldface lower-case letters and numerals will denote either vectors or column matrices, according to the context.[2]

§3. MIXED STRATEGIES

Let S_r be the set of r-dimensional vectors $\vec{\alpha}$ for which $\vec{\alpha} \cdot \vec{1} = 1$ and $\alpha_k \geq 0$, $k = 1, 2, \ldots, r$. S_r is a closed, convex hyperpolyhedron with the unit vectors for vertices.

A <u>mixed strategy</u> for the player I, in which he plays each strategy i with a probability $\xi_i \geq 0$, will be represented by the

[1]Received July 20, 1948 by the ANNALS OF MATHEMATICS and accepted for publication; transferred by mutual consent to ANNALS OF MATHEMATICS STUDY No. 24.

Douglas Aircraft Company, Inc., Project RAND. The authors wish to acknowledge the active assistance of Professor M. A. Girshick of Stanford University, a RAND consultant.

[2]In this typescript version, each letter or numeral carrying a superior dot or arrow, as well as each instance of the upper-case letter "M," should be construed as boldface.

vector $\vec{\xi} \in S_m$. Similarly we let $\vec{\eta} \in S_n$ represent a mixed strategy for player II. It might be noted that the strategies themselves correspond to the vertices of S_m and S_n respectively.

The expectation of player I, when both employ mixed strategies, is

$$\phi(\vec{\xi}, \vec{\eta}) = \Sigma \Sigma\, a^i_j \xi_i \eta_j = \vec{\xi}^T M \vec{\eta}.$$

A convex linear combination of a set of mixed strategies $\vec{\xi}^{(1)}, \vec{\xi}^{(2)}, \ldots, \vec{\xi}^{(r)}$ is any mixed strategy $\vec{\xi}$ of the form:

$$\vec{\xi} = \sum_1^r \alpha_h \vec{\xi}^{(h)}, \quad \text{with } \vec{\alpha} \in S_r .$$

§4. SOLUTIONS

A solution of M (or of Γ) is a pair $\vec{x}, \vec{y}$ of mixed strategies for which

$$(1) \qquad \max_{\vec{\xi} \in S_m} \phi(\vec{\xi}, \vec{y}) = \min_{\vec{\eta} \in S_n} \phi(\vec{x}, \vec{\eta}) = \phi(\vec{x}, \vec{y}) ,$$

or equivalently (see [4], p. 151):

$$(2) \qquad \max_i \vec{a}^{\,i} \cdot \vec{y} = \min_j \vec{a}_j \cdot \vec{x} = \phi(\vec{x}, \vec{y}) .$$

For every solution $\vec{x}, \vec{y}, \phi(\vec{x}, \vec{y})$ must have the same value. This is called the value of the game (to player I) and designated by $v(M)$ or simply v. It should be noted that the game Γ' formed from Γ by adding a constant b to each element of M has the same set of solutions as Γ and a value $v(M') = v(M) + b$. Since we may, by this means, assure a non-zero value by making all the $a^i_j > 0$, in the search for solutions the assumption $v \neq 0$ implies no loss of generality.

A simple solution is one for which

$$\vec{a}^{\,i} \cdot \vec{y} = \vec{a}_j \cdot \vec{x} = \phi(\vec{x}, \vec{y}) = v$$

for all i and j.

A mixed strategy $\vec{\xi}$ or $\vec{\eta}$ is called optimal if it belongs to some solution. It follows from equation (1) that any pair of optimal mixed strategies must constitute a solution of M. We therefore turn our attention to the sets $X \subseteq S_m$ and $Y \subseteq S_n$ of optimal mixed strategies for the two players.

THEOREM 1. X and Y are bounded, non-null, convex, and closed.

PROOF. (a). Bounded: S_r has the diameter $\sqrt{2}$. As subsets of S_m and S_n, X and Y must therefore be bounded.

(b). Non-null: That every game has a solution, and hence an optimal mixed strategy for each player, is proved in [4], Chapter III, Section 17. More general theorems, including this result or important to its proof, have been given by von Neumann [3], Ville [5], Loomis [2], and Wald [6].

(c). Convex: Let

$$\vec{\xi}_0 = \sum_{h=1}^{r} \alpha_h \vec{x}^{(h)} ,$$

where $\vec{\alpha} \in S_r$, $\vec{x}^{(h)} \in X \subseteq S_m$. Since S_m is convex, $\vec{\xi}_0 \in S_m$. Take $\vec{y} \in Y$, then

$$\max_{\vec{\xi}} \phi(\vec{\xi}, \vec{y}) = v .$$

Also

$$\min_{\vec{\eta}} \phi(\vec{\xi}_0, \vec{\eta}) \leq \max_{\vec{\xi}} \min_{\vec{\eta}} \phi(\vec{\xi}, \vec{\eta}) = v$$

and

$$\min_{\vec{\eta}} \phi(\vec{\xi}_0, \vec{\eta}) = \min_{\vec{\eta}} \Sigma_h \alpha_h \phi(\vec{x}^{(h)}, \vec{\eta}) \geq \Sigma_h \alpha_h \min_{\vec{\eta}} \phi(\vec{x}^{(h)}, \vec{\eta}) = v .$$

Therefore,

$$\max_{\vec{\xi}} \phi(\vec{\xi}, \vec{y}) = \min_{\vec{\eta}} \phi(\vec{\xi}_0, \vec{\eta}) = v ,$$

and

$$\vec{\xi}_0 \in X .$$

(d). Closed: Let

$$\vec{\xi}_0 = \lim_{h \to \infty} \vec{x}^{(h)} ,$$

where $\vec{x}^{(h)} \in X \subseteq S_m$, $h = 1, 2, \ldots$. Since S_m is closed, $\vec{\xi}_0 \in S_m$. Taking $\vec{y} \in Y$, as before,

$$\max_{\vec{\xi}} \phi(\vec{\xi}, \vec{y}) = v .$$

Also,

$$\min_{\vec{\eta}} \phi(\vec{\xi}_0, \vec{\eta}) \leq \max_{\vec{\xi}} \min_{\vec{\eta}} \phi(\vec{\xi}, \vec{\eta}) = v ,$$

and

$$\min_{\vec{\eta}} \phi(\vec{\xi}_0, \vec{\eta}) = \min_{\vec{\eta}} \lim_h \phi(\vec{x}^{(h)}, \vec{\eta})$$

due to the linearity of the function ϕ. Then, since $\phi(\vec{x}^{(h)}, \vec{\eta}) \geq v$ for all h and $\vec{\eta}$,

$$\min_{\vec{\eta}} \lim_h \phi(\vec{x}^{(h)}, \vec{\eta}) \geq v .$$

Therefore,

$$\max_{\vec{\xi}} \phi(\vec{\xi}, \vec{y}) = \min_{\vec{\eta}} \phi(\vec{\xi}_o, \vec{y}) = v ,$$

and

$$\vec{\xi}_o \in X .$$

A direct consequence of the convexity is that any game which has more than one solution has infinitely many.

§5. BASIC SOLUTIONS

Given a bounded, non-null, convex, and closed set Z in r-space, we will define Z^* as that subset of Z which contains a vector $\vec{z}^*$ if and only if there do not exist two distinct vectors $\vec{z}'$ and $\vec{z}''$ of Z such that

$$(3) \qquad \vec{z}^* = \tfrac{1}{2}(\vec{z}' + \vec{z}'') .$$

In consequence of the properties of Z, Z^* is non-null and is the smallest set whose convex hull is Z, i.e., which "spans" Z in the sense of convex linear combinations. A <u>basic solution</u> will be defined as the vector pair $\vec{x}^*, \vec{y}^*$ where $\vec{x}^* \in X^*$ and $\vec{y}^* \in Y^*$. The set of all basic solutions thus completely characterizes the solutions of M, i.e., $\vec{\xi}, \vec{\eta}$ is a solution of M if and only if $\vec{\xi}$ and $\vec{\eta}$ are convex linear combinations of the sets of basic optimal mixed strategies X^* and Y^* respectively.

We introduce now the following notation: If $\dot{M}$ is a sub-matrix of M, then $\dot{x}$ and $\dot{a}_j$ ($\dot{y}$ and $\dot{a}^i$) represent the contractions of the vectors $\vec{x}$ and $\vec{a}_j$ ($\vec{y}$ and $\vec{a}^i$) obtained by deleting those components corresponding to the rows (columns) of M which were suppressed in $\dot{M}$. $\dot{\Sigma}_j$ and $\dot{\Sigma}_i$ will denote the analogous contractions of

$$\sum_{j=1}^{n} \quad \text{and} \quad \sum_{i=1}^{m} .$$

THEOREM 2. Given $\vec{x} \in X$, $\vec{y} \in Y$, $v(M) \neq 0$. A necessary and sufficient condition that $\vec{x} \in X^*$ and $\vec{y} \in Y^*$ is that there exist a non-singular sub-matrix $\dot{M}$ of M for which

$$\dot{x}^T = \frac{\dot{1}^T \dot{M}^{-1}}{\dot{1}^T \dot{M}^{-1} \dot{1}} ,$$

(4) $$\dot{y} = \frac{\dot{M}^{-1} \dot{1}}{\dot{1}^T \dot{M}^{-1} \dot{1}} ,$$

$$v = \frac{1}{\dot{1}^T \dot{M}^{-1} \dot{1}} .$$

PROOF. (a). Preliminary: Let M_1 be the sub-matrix which suppresses precisely those rows and columns of M that correspond to the components of $\vec{x}$ and $\vec{y}$ which are zero. Any $\dot{M}$ satisfying (4) must contain M_1 because (4) requires that $\dot{\Sigma}_i x_i = \dot{\Sigma}_j y_j = 1$. It therefore follows that (4) further requires $\dot{x}, \dot{y}$ to be a simple solution of $\dot{M}$, and that $v(\dot{M}) = v(M)$. Now let M_2 be the sub-matrix which suppresses precisely those rows and columns of M that correspond to the components of $M\vec{y}$ and $M^T\vec{x}$ which are different from $v(M)$. Then $\dot{M}$ can satisfy (4) only if

(5) $$M_1 \subseteq \dot{M} \subseteq M_2 .$$

(b). Sufficiency: Suppose $\vec{x} \notin X^*$, and take any $\dot{M}$ satisfying (5). Then

(6) $$\dot{x}^T \dot{M} = v\dot{1}^T .$$

By (3) there exist two distinct optimal mixed strategies $\vec{x}\,'$ and $\vec{x}\,''$ such that $\vec{x} = \frac{1}{2}(\vec{x}\,' + \vec{x}\,'')$. Obviously $\dot{x} = \frac{1}{2}(\dot{x}\,' + \dot{x}\,'')$, where the components deleted in forming $\dot{x}'$ and $\dot{x}''$ are all zero. Then, if $\dot{a}_j$ is a column of $\dot{M}$,[3]

$$\dot{x}' \cdot \dot{a}_j = \vec{x}\,' \cdot \vec{a}_j \geq v, \quad \text{and} \quad \dot{x}'' \cdot \dot{a}_j = \vec{x}\,'' \cdot \vec{a}_j \geq v ,$$

while, by (6),

$$\tfrac{1}{2}(\dot{x}' + \dot{x}'') \cdot \dot{a}_j = v .$$

Therefore:

$$(\dot{x}')^T \dot{M} = v\dot{1}^T = (\dot{x}'')^T \dot{M} ;$$

or

$$(\dot{x}' - \dot{x}'')^T \dot{M} = \dot{0} .$$

But, since $\vec{x}\,' \neq \vec{x}\,''$, this means that $\dot{M}$ is singular.

[3]Observe that $\dot{a}_j$ is defined for every j between 1 and n; the dot shortens the vector but does not reduce the range of the index j.

If $\vec{y} \in Y^*$, the argument is exactly similar.

(c). Necessity: Suppose $\vec{x} \in X^*$ and $\vec{y} \in Y^*$. We shall exhibit a sub-matrix with the desired properties. Starting with M_1, adjoin to it certain rows and columns as follows: every row (of the same length as the rows of M_1), contained in M_2 but not M_1, is considered in turn and is <u>adjoined</u> or <u>discarded</u> according as it <u>is</u> or <u>is not</u> linearly independent of the set consisting of the rows of M_1 and the previously adjoined rows. Every column (again of the same length as the columns of M_1), in M_2 but not M_1, is adjoined or discarded on the same grounds. The smallest matrix containing M_1 and the adjoined rows and columns will be called a <u>kernel</u> of the basic solution $\vec{x}$, $\vec{y}$. (Since the order of "consideration" is arbitrary, a kernel of a basic solution is not necessarily unique.)

Let $\dot{M}_0$ be a kernel of $\vec{x}$, $\vec{y}$ and suppose that $\dot{M}_0$ is singular. Since the construction of $\dot{M}_0$ was symmetrical in regard to rows and columns, we may assume that the rows of $\dot{M}_0$ form a dependent set. The dependency will not involve the adjoined rows.[4] Therefore there exists a set of constants $\vec{c}$: $(c_1, c_2, \ldots, c_m)$ not all zero such that

$$\dot{\Sigma}_i \, c_i \dot{a}^i = \dot{0} \tag{7}$$

and

$$c_i = 0 \quad \text{if} \quad x_i = 0 \ . \tag{8}$$

Now, if $\dot{a}^i$ is a row of $\dot{M}_0$,

$$\dot{y} \cdot \dot{a}^i = v \ .$$

Therefore,

$$\dot{\Sigma}_i \, c_i v = \dot{\Sigma}_i \, c_i \dot{y} \cdot \dot{a}^i = \dot{y} \cdot (\dot{\Sigma}_i \, c_i \dot{a}^i) = \dot{y} \cdot \dot{0} = 0 \ ,$$

or, since $v \neq 0$,

$$\dot{\Sigma}_i \, c_i = \sum_1^m c_i = 0 \ . \tag{9}$$

Consider the vectors $\vec{x} \pm \varepsilon \vec{c}$. For $|\varepsilon|$ sufficiently small, by (8) and (9),

$$\vec{x} \pm \varepsilon \vec{c} \in S_m \ .$$

Further, by (2), $\vec{x} \pm \varepsilon \vec{c} \in X$, if

$$\min_j (\vec{x} \pm \varepsilon \vec{c}) \cdot \vec{a}_j = v \ . \tag{10}$$

---

[4] The construction of $\dot{M}_0$ dealt with sub-rows. But if contracted vectors are independent, then their extensions must be independent as well.

Now, by (7), $\vec{c} \cdot \vec{a}_j = \dot{c} \cdot \dot{a}_j = 0$ for all $\dot{a}_j$ in $\dot{M}_0$. Therefore $\vec{c} \cdot \vec{a}_j = 0$ for all $\dot{a}_j$ outside $\dot{M}_0$ but "in" M_2 by the dependency condition used in constructing $\dot{M}_0$. Thus, for all $\dot{a}_j$ "in" M_2,

$$(\vec{x} \pm \varepsilon \vec{c}) \cdot \vec{a}_j = \vec{x} \cdot \vec{a}_j \pm \varepsilon \vec{c} \cdot \vec{a}_j = v .$$

However, for every $\dot{a}_j$ "outside" M_2, $\vec{x} \cdot \vec{a}_j > v$. There exists, therefore, an $\varepsilon > 0$ such that (10) holds. However, the existence in X of the vectors $\vec{x} + \varepsilon \vec{c}$ and $\vec{x} - \varepsilon \vec{c}$ contradicts the hypothesis that $\vec{x} \in X^*$. Thus $\dot{M}_0$ cannot be singular.

Now by construction,

$$\dot{x}^T \dot{M}_0 = v \dot{1}^T , \quad \dot{M}_0 \dot{y} = v \dot{1} .$$

So, since $\dot{M}_0$ is non-singular,

$$\dot{x}^T = v \dot{1}^T \dot{M}_0^{-1} , \quad \dot{y} = v \dot{M}_0^{-1} \dot{1} .$$

But since

$$v \dot{1}^T \dot{M}_0^{-1} \dot{1} = \dot{x}^T \cdot \dot{1} = 1 ,$$

equations (4) follow directly.

COROLLARY 1.[5] A necessary and sufficient condition that a solution $\vec{x}, \vec{y}$ of M be basic is that there exist a square sub-matrix $\dot{M}$ of M such that[6]

$$\dot{x}^T = \frac{\dot{1}^T \operatorname{adj} \dot{M}}{\dot{1}^T \operatorname{adj} \dot{M} \dot{1}}$$

(11)
$$\dot{y} = \frac{\operatorname{adj} \dot{M} \dot{1}}{\dot{1}^T \operatorname{adj} \dot{M} \dot{1}}$$

$$v(M) = \frac{|\dot{M}|}{\dot{1}^T \operatorname{adj} \dot{M} \dot{1}}$$

---

[5]The formula for $v(M)$, as given in (11), has been derived by Kaplansky in [1] for the special case of a "completely mixed" game, i.e., a game whose optimal mixed strategies contain no zero components.

[6]Formulae (11) are considered not to hold if the right-hand members are indeterminate. This interpretation is necessary because of the fact that the accumulation points of the set of basic solutions of a convergent sequence of matrices are not necessarily all basic solutions of the limiting matrix. Thus (11) may be true in the limit even for some non-basic $\vec{x}, \vec{y}$; but only with the aid of an $\dot{M}$ which is singular in the limit.

PROOF. (a). For $v(M) \neq 0$, observe that the formula for $v(M)$ in (11) holds only if $\dot{M}$ is non-singular; but in that case the corollary may be derived immediately from the theorem by means of the relation $\text{adj } \dot{M} = |\dot{M}|\dot{M}^{-1}$.

(b). For $v(M) = 0$, form the game Γ' by adding a constant $b \neq 0$ to each element of M, i.e.

$$M' = ||a^i_j + b|| \quad (12)$$

Then,

$$v(M') = v(M) + b \neq 0$$

and

$$X', Y', X'^*, Y'^* = X, Y, X^*, Y^* \quad \text{(respectively)} .$$

By the case already proved, a necessary and sufficient condition that $\vec{x}$, $\vec{y}$ be a basic solution of M' and hence of M, is the existence of $\dot{M}'$ such that

$$\dot{x}^T = \frac{\dot{1}^T \text{adj } \dot{M}'}{\dot{1}^T \text{adj } \dot{M}' \dot{1}}$$

$$\dot{y} = \frac{\text{adj } \dot{M}' \dot{1}}{\dot{1}^T \text{adj } \dot{M}' \dot{1}} \quad (13)$$

$$v(M) + b = \frac{|\dot{M}'|}{\dot{1}^T \text{adj } \dot{M}' \dot{1}} .$$

By means of some elementary properties of determinants[7], the expression (13) may be reduced to (11), and hence the corollary is proved.

COROLLARY 2. Every game Γ has a finite number of basic solutions, and the sets X and Y of optimal mixed strategies are hyper-polyhedra.

---

[7]The key relationships, both following from (12), are

$$\dot{1}^T \text{adj } \dot{M}' = \dot{1}^T \text{adj } \dot{M}$$

and

$$|\dot{M}'| = |\dot{M}| + b\, \dot{1}^T \text{adj } \dot{M}\, \dot{1} .$$

The latter is most easily proved by considering the derivatives of $|\dot{M}'|$ with respect to b, of which all except the first vanish identically.

PROOF. (a). An inspection of (11) discloses that a sub-matrix can be the kernel of at most one basic solution of Γ. But every basic solution has at least one kernel. Hence there cannot be more basic solutions of Γ than there are sub-matrices of M.

(b). X^* and Y^* are both finite sets of vectors, else the number of basic solutions would be infinite. But the convex hull of a finite set is a hyper-polyhedron (whose dimension is less than the number of points in the set).

§6. CONSTRUCTIVE METHOD FOR DETERMINING ALL SOLUTIONS

As has been pointed out, every basic solution is associated with at least one square sub-matrix (basic kernel). Therefore, to find all basic solutions it is sufficient to consider all square sub-matrices in the light of Corollary 1. A systematic procedure would be to consider all r x r sub-matrices for $r = 1, 2, .., \min(m, n)$. To each r x r sub-matrix $\dot{M}$, apply formulae (11) and determine whether $\dot{x}$ and $\dot{y}$ belong to S_r. If not, that sub-matrix cannot be a kernel; if so, determine by (2) whether the extended vectors $\vec{x}$ and $\vec{y}$ belong to X and Y respectively. This will be the case if and only if $\vec{x}, \vec{y}$ is a basic solution. This procedure must find every basic solution and will thus determine the set of all solutions.

REFERENCES

[1] KAPLANSKY, I., "A Contribution to von Neumann's Theory of Games," Annals of Mathematics, Vol. 46 (1945), pp. 474-479.

[2] LOOMIS, L. H., "On a Theorem of von Neumann," Proceedings of the National Academy of Sciences, Vol. 32 (1946), pp. 213-215.

[3] von NEUMANN, J., "Zur Theorie der Gesellschaftsspiele," Math. Annalen, Vol. 100 (1928), pp. 295-320.

[4] MORGENSTERN, O., von NEUMANN, J., "Theory of Games and Economic Behavior," 2nd ed. (1947), Princeton University Press, Princeton.

[5] VILLE, J., "Sur la Théorie Générale des Jeux où Intervient L'habilité des Joueurs," Traité du Calcul des Probabilités et de ses Applications by Emil Borel and collaborators, Vol. 2, Part 5 (1938), Paris, pp. 105-113.

[6] WALD, A., "Foundations of a General Theory of Sequential Decision Functions," Econometrica, Vol. 15 (1947), pp. 279-313.

L. S. Shapley

R. N. Snow

The RAND Corporation

SOLUTIONS OF FINITE TWO-PERSON GAMES[1] *

D. Gale and S. Sherman

The fundamental theorem of the finite two-person game asserts the existence of a value and optimal mixed strategies for both players in any game Γ. The optimal strategies need not be unique. It is easy to show, however, that the optimal strategies for an m by n game form in a natural way polyhedral subsets of an $(m - 1)$- and $(n - 1)$-dimensional simplex respectively, but that not all such pairs of polyhedra can be obtained as solutions of a game. The purpose of this paper is two-fold.

1. To give a simple characterization of the sets of optimal strategies which can occur as solutions of an m by n game.

2. Given such sets to characterize all games having these sets as their optimal strategies.

The principal tool to be used in this investigation will be the theory of polyhedral convex cones, in particular the treatment of the subject given by H. Weyl [1]. This technique has the advantage that the proofs given are completely algebraic; that is, no use is made of the completeness of the real numbers or of continuity arguments. The whole treatment, therefore, carries over at once to any ordered field (an observation which has been made by H. Weyl), in particular to games with the rationals replacing the reals as basic field. Theorems which have been proved elsewhere by topological methods are reproved here in purely algebraic fashion.

Let us recall briefly the notions of the two-person game. We are given an m by n matrix A which may be looked on as giving a transformation from an n-dimensional vector space V associated with the second player into an m-dimensional space U associated with the first player. The unit vectors of U and V which correspond to the pure strategies of the game associated with A are denoted by $u_1, \ldots, u_m$ and $v_1, \ldots, v_n$. The set of mixed strategies for the two players consist of all positive linear combinations of the unit vectors such that the sum of the components is 1. The sets of such vectors form $(m - 1)$- or $(n - 1)$-dimensional simplexes, denoted by Q_1 or Q_2. The game is said to have value φ if there exist vectors (mixed strategies) $x_0 \in Q_1$ and $y_0 \in Q_2$ such that for any $v \in Q_2$ and $u \in Q_1$ we have

[1] This work was supported by Office of Naval Research funds.

*Accepted as a direct contribution to ANNALS OF MATHEMATICS STUDY No. 24.

$$x_0 Av \geq \varphi \geq uAy_0 ,$$

where the expressions x_0Av and uAy_0 denote matrix multiplication, where x_0 and u are considered to be 1 by m matrices and v and y_0 are n by 1 matrices. The game theoretic interpretation of the above is assumed known and will not be described here.

A mixed strategy $x \in Q_1$ is called optimal if $xAv \geq \varphi$ for all $v \in Q_2$. Similarly, $y \in Q_2$ is optimal if $uAy \leq \varphi$ for all $u \in Q_1$. The sets of all optimal strategies for first and second players will be denoted by X and Y respectively. The fundamental theorem of the two-person game asserts that X and Y are non-empty. As already stated, our purpose will be to give a complete analysis of these sets. The principal result is given in paragraph 2, but it can be described here roughly as follows: The sets X and Y are convex polyhedra lying in the simplexes Q_1 and Q_2. If we define the carriers of X and Y to be the faces of the simplexes of smallest dimension containing them, denoted by E_1 and E_2, then our first condition on X and Y is:

$$\text{dimension } E_1 - \text{dimension } X = \text{dimension } E_2 - \text{dimension } Y.$$

This relation was first proved by Bohnenblust, Karlin, and Shapley [2].

The second condition relates the number of faces of the polyhedra X and Y to the number of vertices of Q_2 and Q_1 respectively. The two conditions completely describe all solutions of games.

In the first paragraph we introduce the essential definitions and facts from the theory of polyhedral cones. The main theorem is then stated and proved in the remaining paragraphs, making use of the cone theory.

§ 1. POLYHEDRAL CONES

A subset C of a finite dimensional (real) vector space V is called a <u>convex cone</u> if,

$$x_1, x_2 \in C \text{ and } \lambda_1, \lambda_2 \geq 0 \text{ implies } \lambda_1 x + \lambda_2 x_2 \in C .$$

If C_1 and C_2 are convex cones then so are the <u>intersection</u> $C_1 \cap C_2$ and the <u>sum</u> $C_1 + C_2$, where

$$C_1 + C_2 = \{x \mid x = x_1 + x_2 \text{ where } x_1 \in C_1, x_2 \in C_2\} .$$

If C is a cone we define $[C]$ to be the smallest linear space containing C, or what is the same thing,

$$[C] = \{v \mid v = x - x', x, x' \in C\} .$$

The <u>dimension</u> of C is defined to be the dimension of $[C]$.

If C is a cone we define

$$C^{\perp} = \{v \mid v \cdot x = 0 \text{ for all } x \in C\} ,$$

where $v \cdot x$ denotes the scalar product of v and x. ($C^{\perp}$ may be read as "C perpendicular.")

The most important notion related to cones is that of polar cone which we now define. If C is a cone its <u>polar cone</u> C^* is defined by

$$C^* = \{v \mid v \cdot x \geq 0 \text{ for all } x \in C\} .$$

Geometrically C^* may be thought of as the set of all vectors making a non-obtuse angle with the vectors of C. One easily verifies the relations,

$$a^*. \quad C_1 \subset C_2 \longrightarrow C_2^* \subset C_1^*$$

$$b^*. \quad (C_1 + C_2)^* = C_1^* \cap C_2^* .$$

If a cone is generated by a single vector x_o, that is $C = \{x \mid x = \lambda x_o, \ \lambda \geq 0\}$, it is called a <u>ray</u> or <u>halfline</u>.

The polar cone $(x_o)^*$ of the halfline (x_o) is called a <u>half-space</u> and is seen to be given by

$$(x_o)^* = \{y \mid y \cdot x_o \geq 0\} .$$

We now give two definitions of a polyhedral cone.

i. <u>Intersection definition.</u> C is a polyhedral cone if it is the intersection of a finite number of halfspaces;

$$C = \bigcap_{j=1}^{l} (y_j)^* .$$

ii. <u>Sum definition.</u> C is a polyhedral cone if it is the sum of a finite number of halflines;

$$C = \sum_{i=1}^{k} (x_i) .$$

The main fact about polyhedral cones is that these two definitions are equivalent. This intuitively obvious fact is by no means trivial to prove. A completely algebraic proof is given by Weyl [1]. His main theorem shows that an n dimensional polyhedral cone C in n-space defined by the sum definition is also a finite intersection of halfspaces. The condition that C be n dimensional is easily removed by first representing $[C]$ as an intersection of halfspaces and then applying Weyl's theorem inside of $[C]$. The fact that an "intersection" cone is also a "sum" cone is a consequence of Theorems 3 and 4 of Weyl's paper. The other facts listed in this paragraph are either proved in [1] or can be obtained as easy corollaries of Weyl's results. For this reason and also because the statements are quite obvious intuitively proofs will not be given here.

We let S be the set of all polyhedral cones in some vector space V and list the following properties:

Property 1. S is closed under the operations $+$, $\cap$ and $*$ (follows at once by using the appropriate definition i or ii of polyhedral cone and property b^*).

Property 2. $C_1 \subset C_2 \longrightarrow C_2^* \subset C_1^*$.

Property 3. $(C_1 + C_2)^* = C_1^* \cap C_2^*$. (Properties 2 and 3 are simply a^* and b^* repeated.)

Property 4. $(C_1 \cap C_2)^* = C_1^* + C_2^*$.

Property 5. $(C^*)^* = C$.

Property 5 is essentially Theorem 10 of Weyl's, while Property 4 is a consequence of Properties 5 and 3.

Sets satisfying the above properties have been discussed in the literature [3] and are called orthocomplemented lattices with respect to the three operations involved. Some further properties of cones will be needed.

DEFINITION. If C is a cone and $y \in C^*$ we define $(y)^\perp$ to be a <u>supporting hyperplane</u> of C and $(y)^\perp \cap C$ to be a <u>face</u> of C.

If C has dimension p and $(y)^\perp \cap C$ has dimension $(p - 1)$ then $(y)^*$ is called an <u>extreme halfspace</u> of C and $(y)^\perp \cap C$ is called a <u>bounding face</u> of C.

Property 6. If C is a polyhedral cone it is the intersection of its extreme halfspaces.

This is a direct corollary to Weyl's main theorem. One final fact will be needed:

LEMMA 1. If $C = \bigcap_{j=1}^{l} (y_j)^*$ and F is a bounding face of C, then there exists y_m, $1 \leq m \leq l$ such that $F = (y_m)^\perp \cap C$.

PROOF. By definition of a face, $F = (y)^\perp \cap C$ for some

$$y \in C^* = \sum_{j=1}^{l} (y_j) \quad \text{so} \quad y = \sum_{j=1}^{l} \lambda_j y_j,\ \lambda_j \geq 0 .$$

Since $\dim F = \dim C - 1$, there exists $\bar{x} \in C$ such that

$$\bar{x} \cdot y = \sum_{j=1}^{l} \lambda_j(\bar{x} \cdot y_j) > 0$$

and hence for some j, say $j = m$, we must have $\lambda_m > 0$ and $(\bar{x} \cdot y_m) > 0$, so $\bar{x} \notin (y_m)^{\perp}$, and therefore $C \not\subset (y_m)^{\perp}$. Let $G = (y_m)^{\perp} \cap C$ and let us show that $G = F$. $F \subset G$, for if $x \in F$ then $x \cdot y = \sum \lambda_j(x \cdot y_j) = 0$ hence $\lambda_j(x \cdot y_j) = 0$ for all j so $(x \cdot y_m) = 0$ since $\lambda_m > 0$. $G \subset F$ for suppose $x' \in G$ but $x' \notin F$. Then $(x') + F$ has the dimension of C (since F is a bounding face) and hence $[x' + F] = [C] \subset (y_m)^{\perp}$ whereas we have shown that $C \not\subset (y_m)^{\perp}$.

§ 2. STATEMENT OF PRINCIPAL THEOREM

For convenience in discussing sets of optimal strategies we make some simplifications of the problem. We first observe that no generality is lost if we consider only games whose value φ is zero as far as the nature of the sets of optimal strategies is concerned, for suppose sets X and Y are the solutions of some game A with value $\varphi \neq 0$. One verifies at once that X and Y are also the solutions of the game whose matrix $\bar{A}$ is obtained from A by subtracting the amount φ from each entry in A, and clearly $\bar{A}$ has value zero. Henceforth, therefore, we restrict ourselves to games with value zero.

The second simplification is the following. We drop the requirement on a mixed strategy that the sum of the components be 1. The set of mixed strategies then becomes the set of all non-negative vectors in U and V (instead of the simplexes) and will be denoted by P_1 and P_2.

With these simplifications the sets X and Y are now redefined as follows

$$X = \{x \mid x \in P_1 \text{ and } xAv \geq 0 \text{ for all } v \in P_2\}$$
$$Y = \{y \mid y \in P_2 \text{ and } uAy \leq 0 \text{ for all } u \in P_1\} .$$

To get from these sets X and Y to the actual sets of optimal strategies one merely takes the intersections of X and Y with the hyperplane through the unit vectors in U and V. Going in the other direction the sets X and Y are simply the cones subtended by the optimal strategies at the origin.

We observe,

LEMMA 2. The sets X and Y are polyhedral cones. (Hence optimal strategies are polyhedra.)

PROOF. Using the notation of the previous section we see that $X = P_1 \cap (AP_2)^*$ (where AP_2 denotes the image of P_2 under the transformation A). Since

$$P_2 = \sum_{j=1}^{n} (v_j) \quad \text{it follows that} \quad AP_2 = \sum_{j=1}^{n} (Av_j)$$

and is therefore a polyhedral cone. Therefore $(AP_2)^*$ is also polyhedral, as is its intersection with P_1, from properties of cones of the previous paragraph. Similarly for the set Y.

Since X and Y are polyhedral cones we may now talk about dimension, bounding faces, etc. Our purpose is to find conditions on these quantities so that X and Y will be solutions of some game. We assume therefore that we are given X and Y in P_1 and P_2 respectively.

DEFINITION. The _essential face_, E_1 of P_1 is defined to be the face of P_1 of lowest dimension which contains X. The unit vectors u_i which span E_1 are called essential unit vectors. Algebraically the vector u_i is essential if there exists $x \in X$ such that $x \cdot u_i > 0$. If we think of X as solutions of a game then u_i is called an _essential pure strategy_ and has the property that it is actually used in some optimal mixed strategy.

If u_i is not essential it is called superfluous. We denote by e_1 and s_1 the number of essential and superfluous unit vectors (pure strategies) in P_1.

In like manner we define E_2 in P_2 and the numbers e_2 and s_2 associated with the set Y.

A bounding face of X or Y will be called _interior_ if it does not lie in some proper subface of E_1 or E_2. We denote by f_1 and f_2 the number of interior bounding faces of X and Y respectively.

Finally let d_1 and d_2 be the dimensions of X and Y.

PRINCIPAL THEOREM. The pair of polyhedral cones X and Y are solutions of a two-person game if and only if,

(1) $e_1 - d_1 = e_2 - d_2$

(2) $f_1 \leq s_2, \; f_2 \leq s_1$.

The remaining paragraphs are concerned with the proof and consequences of this theorem.

§3. PROOF OF NECESSITY

In this paragraph we prove the necessity of conditions (1) and (2) of the main theorem. The proof of (1) was first given by Bohnenblust, Karlin, and Shapley in [2] and makes use of a compactness argument. We give our own proof here which is purely algebraic.

Condition (1) is clearly equivalent to the following:

$$\dim E_1 - \dim X = \dim E_2 - \dim Y .$$

We adopt the following notation. If S is a subset of V then AS denotes the image of S under the transformation A from V to U, and if R is a subset of U the RA is the image of R in V under the transformation A transpose.

LEMMA 3. $X^* = P_1 + AP_2$, $Y^* = P_2 - P_1A$.

PROOF. As seen in Paragraph 2 $X = P_1 \cap (AP_2)^*$ (Lemma 2). Now

$$P_1 = \bigcap_{i=1}^{m} (u_i)^* = (\sum_{i=1}^{m} u_i)^* = P_1^* \quad \text{so} \quad X = P_1^* \cap (AP_2)^* = (P_1 + AP_2)^*$$

by Property 3 of cones. Hence $X^* = P_1 + AP_2$ by Property 5, and similarly $Y^* = P_2 + (-P_1A) = P_2 - P_1A$.

LEMMA 4. The pure strategy u_{i_o} is essential if and only if $u_{i_o}Ay = 0$ for all $y \in Y$.

PROOF. Suppose u_i is essential. Then there exists $x \in X$ for which $x \cdot u_{i_o} > 0$. Now

$$x = \sum_{i=1}^{m} (x \cdot u_i)u_i \quad \text{and} \quad xAy = 0 \quad \text{for all} \quad y \in Y \quad \text{so} \quad \sum_{i=1}^{m} (x \cdot u_i)(u_iAy) = 0$$

and since each term is non-positive we get $(x \cdot u_i)(u_iAy) = 0$ for all i, so since $x \cdot u_{i_o} > 0$ it follows that $u_{i_o}Ay = 0$.

To prove the converse we show that if u_{i_o} is superfluous then there exists $y \in Y$ such that $u_{i_o}Ay < 0$. First $u_{i_o} \in X^{\perp}$ so $-u_{i_o} \in X^{\perp} \subset X^* = P_1 + AP_2$ by the previous lemma. Hence $-u_{i_o} = u + Av$, $u \in P_1$, $v \in P_2$.

Now

(a) for $i \neq i_o$ $0 = u_i \cdot u + u_iAv$, so $u_iAv = -u_i \cdot u \leq 0$

(b) for $i = i_o$ $-1 = u_{i_o} \cdot u + u_{i_o}Av$, so $u_{i_o}Av = -1 - u_{i_o} \cdot u < 0$.

Thus (a) and (b) together show that v is in Y, and from (b) alone $u_{i_0} Av < 0$ as was to be shown.

The above lemma is also proved in [2] by different methods. From here on, however, in obtaining the necessity of condition (1) we parallel quite closely the procedure of [2], the proofs being given here for the sake of completeness.

LEMMA 5. There exists $y_0 \in Y$ such that $y_0 \cdot v_j > 0$ for all essential v_j and $u_i A y_0 < 0$ for all superfluous u_i.

PROOF. For each essential v_j there exists by definition $y_j \in Y$ such that $y_j \cdot v_j > 0$. Also by the previous lemma for each superfluous u_i there exists $y_i' \in Y$ such that $u_i A y_i' < 0$. Defining

$$y_0 = \sum_{j=1}^{e_2} y_j + \sum_{i=1}^{s_1} y_i'$$

it is clear that the conditions of the lemma are satisfied.

DEFINITION. Let A_e be the submatrix of A whose rows and columns correspond to the essential unit vectors of U and V. This is the essential submatrix or subgame. The matrices A_e and A_e' (A_e transpose) induce transformations between $[E_1]$ and $[E_2]$. The null space of these transformations will be denoted by $N(A_e)$ and $N(A_e')$.

THEOREM 1. $[Y] = N(A_e)$, $[X] = N(A_e')$.

PROOF. For $v \in [Y]$ we have $v = y - y'$ where $y, y' \in Y$ and for u_i essential we get $u_i A_e v = u_i A v = u_i A y - u_i A y' = 0$ from Lemma 4, and hence $v \in N(A_e)$.

Conversely suppose that $v \in N(A_e)$, that is $A_e v = 0$. Choose y_0 satisfying the conditions of Lemma 5. Now define y' to be $v + \lambda y_0$ where λ is chosen so large that

(a) $(v + \lambda y_0) \cdot v_j \geq 0$ for $v_j \in E_2$

(b) $u_i A(v + \lambda y_0) = u_i A v + \lambda(u_i A y_0) \leq 0$ for $u_i \in S_1$

(S_1 and S_2 denote the cones spanned by the superfluous strategies in U and V respectively).

Note also that

(c) $(v + \lambda y_0) \cdot v_j = 0$ for $v_j \in S_2$ since $v \in E_2$

(d) $u_i A_e(v + \lambda y_0) = u_i A_e v + \lambda u_i A_e y_0 = 0$ for $u_i \in E_1$

by assumption.

From (a) and (c) $y' \in P_2$ and from (b) and (d) $y' \in -(P_1A)^*$ hence $y' \in Y$ so $v = y' - \lambda y_0 \in [Y]$. Similarly $N(A_e') = [X]$.

The necessity of condition (1) follows as an immediate consequence of this theorem.

COROLLARY. $\dim E_2 - \dim Y = \dim E_1 - \dim X$.

PROOF. $\dim E_2 - \dim Y = \dim [E_2] - \dim [Y] = \dim [E_2] - \dim N(A_e) = \text{rank } A_e = \text{rank } A_e' = \dim [E_1] - \dim N(A_e') = \dim E_1 - \dim X$.

It remains to show the necessity of condition (2), and it suffices to prove that $f_1 \leq s_2$. To do this write

$$X = \bigcap_{j=1}^{n} (Av_j)^* \bigcap_{i=1}^{m} (u_i)^* .$$

Now suppose F is an interior bounding face of X. Since it is interior it is not of the form $(u_i)^{\perp} \cap X$ and therefore by Lemma 1 of §1,

$$F = (Av_j)^{\perp} \cap X \quad \text{for some } v_j .$$

Further this v_j must be superfluous, for if not then by Lemma 4, $xAv_j = 0$ for all $x \in X$ or $X \subset (Av_j)^{\perp}$, hence $F = X \cap (Av_j)^{\perp} = X$, contradicting the fact that F is a proper face of X.

Thus we have seen that to each interior bounding face F there corresponds a superfluous strategy and it is clear that distinct faces correspond to distinct strategies from which condition (2) follows.

§ 4. PROOF OF SUFFICIENCY

We will now demonstrate the sufficiency of conditions (1) and (2) of the principal theorem. The bulk of the work is done in the following lemma.

LEMMA 6. If X and Y satisfy (1) and (2) then there exist linear transformations A_e of $[E_2]$ into $[E_1]$, A_2 of $[S_2]$ into $[E_1]$ and A_1 of $[S_1]$ into $[E_2]$ such that

(a) $[Y] = N(A_e)$, $[X] = N(A_e')$

(b) $A_2S_2 \subset X^*$ and $A_2S_2 \cap X^{\perp} = 0$

$S_1A_1 \subset -Y^*$ and $S_1A_1 \cap Y^{\perp} = 0$

(c) for each interior bounding face F of x there exists $v_j \in S_2$ such that $(A_2v_j)^{\perp} \cap X = F$.

For each interior bounding face G of Y there exists $u_i \in S_1$ such that $(u_iA_1)^{\perp} \cap Y = G$.

PROOF. To show that we may satisfy (a) merely observe that by condition (1) the quotient space $[E_2]/[Y]$ has the same dimension as the linear space $X^{\perp} \cap [E_1]$ and therefore we can define a linear map A_e from $[E_2]$ onto $X^{\perp} \cap [E_1]$ with null space $[Y]$. Then also $N(A_e') = [X]$ for if $x \in X$, then for any $v \in E_2$, $xA_ev = 0$ since $A_ev \in X^{\perp}$. On the other hand if $u \in N(A_e')$ then $uA_ev = 0$ for all $v \in [E_2]$ so $u \in (A_eE_2)^{\perp} \cap [E_1]$ $= (X^{\perp} \cap [E_1])^{\perp} \cap [E_1] = ((X^{\perp})^{\perp} + E_1^{\perp}) \cap [E_1] = [X]$.

To satisfy (b) and (c) choose for each interior bounding face F_j of X a vector u in $[E_1] \cap X^*$ such that $(u)^{\perp} \cap X = F_j$ and then pick a unit vector $v_j \in S_2$ and define $A_2v_j = u$. This is possible since by condition (2) the dimension of S_2 exceeds the number of interior bounding faces of X. For these v_j clearly $A_2v_j \in X^*$ and since $F_j \neq X$ there exists $x \in X$, $x \notin F_j$ for which $xA_2v_j \neq 0$, and so $A_2v_j \notin X^{\perp}$. If we now define A_2 on the remaining unit vectors of S_2 so that each is mapped into the cone spanned by those for which A_2 has already been defined then (b) and (c) will be satisfied.

From this lemma we quickly obtain the proof of sufficiency and the characterization of all matrices with solutions X and Y.

THEOREM 2. Given a pair of cones X and Y satisfying the conditions of the principal theorem, they are solutions of a matrix A if and only if A can be decomposed as follows:

$$A = \begin{pmatrix} A_e & A_2 \\ A_1 & A_s \end{pmatrix}$$

where A_s is arbitrary and A_e, A_1, A_2 satisfy the conditions (a), (b), (c) of the preceding lemma.

PROOF. We have already shown the existence of such a matrix A. On the other hand given any matrix A we can decompose it as above. We first show that if A has solutions X and Y it must satisfy (a), (b), and (c).

The necessity of (a) is exactly the statement of Theorem 1. To verify the necessity of condition (b) we have by Lemma 3, $X^* = P_1 + AP_2$ so $AP_2 \subset X^*$ and, in particular, $AS_2 \subset X^*$ so $A_2S_2 \subset X^* \cap E_1$. To show that $A_2S_2 \cap X^{\perp} = 0$ notice that if v_j is any unit basis vector in V and $Av_j \in X^{\perp}$ then $xAv_j = 0$ for all $x \in X$ and hence by Lemma 4, v_j is essential. Hence for $v_j \in S_2$ there exists $x \in X$ such that

$xAv_j = xA_2v_j \neq 0$ and therefore $A_2v_j \notin X^{\perp}$. Then for non-zero

$$v = \sum_{j=1}^{s_e} \lambda_j v_j \in S_2$$

we cannot have $xA_2v = 0$ since this would imply $xAv_j = 0$ for some v_j.

For condition (c) we again apply directly Lemma 1 and show that there exists v_j such that $(Av_j)^{\perp} \cap X = F$ for each interior bounding face F of X. This v_j is not essential for if it were Lemma 4 asserts $Av_j \in X^{\perp}$ or $(Av_j)^{\perp} \supset X$ which is not possible. Since $v_j \in S_2$ we obtain $xAv_j = xA_2v_j = 0$ and hence $F = (A_2v_j)^{\perp} \cap X$.

We must finally show that any matrix of the form A has as its solutions precisely the sets X and Y. Designating by $\sigma_1(A)$ the solution of A for player I, we first show $X \subset \sigma_1(A)$. For $x \in X$, $xAv_j = xA_ev_j = 0$ for $v_j \in E_2$, and for $v_j \in S_2$, $xAv_j = xA_2v_j \geq 0$ since $A_2v_j \in X^*$. By a similar argument for $y \in Y$ we see that A has value zero and the sets X and Y are among the optimal strategies.

To show that $\sigma_1(A) \subset X$ we first show that $\sigma_1(A) \subset E_1$, for suppose $u_i \in S_1$. Then $u_iAy = u_iA_1y < 0$ for some $y \in Y$ by condition (b) and therefore by Lemma 4, u_i is superfluous.

To show $\sigma_1(A) \subset X$ we suppose $u \in \sigma_1(A) \subset E_1$. If $v_j \in E_2$ then by Lemma 4, $uAv_j = uA_ev_j = 0$ so $u \in N(A_e') = [X]$ by condition (a) and so $s_1(A) \subset [X]$. Now by Property 6 of cones, X is the intersection of its extreme halfspaces and $[X]$. By condition (c) each such halfspace is of the form $(u_i)^*$ or $(A_2v_j)^*$ for some $v_j \in S_2$ so we need only prove that $uAv_j \geq 0$ for all $v_j \in S_2$, but this follows from condition (b) hence the proof is complete.

§5. INTERPRETATION OF RESULTS AND APPLICATIONS

NORMALIZED MIXED STRATEGIES

We now return to the simplexes of normalized mixed strategies Q_1 and Q_2 and the corresponding sets of optimal strategies. As was pointed out these sets are simply the intersection of X and Y with the planes through the unit vectors in U and V. Thus the statement of the principal theorem still stands with interior bounding faces and dimension suitably reinterpreted.

UNIQUE OPTIMAL STRATEGIES

Applying condition (1) of the main theorem to a very special case gives the interesting fact:

If the optimal strategies are unique for both players then $e_1 = e_2$. This is clear since for this case $d_1 = d_2 = 0$.

Another application is concerned with completely mixed games. Kaplansky [4] has defined a game to be completely mixed if every optimal strategy makes use of all pure strategies. As a result of our theory we can state:

A completely mixed game has unique solutions for both players (hence by the previous application it must be square).

This follows out of condition (2) of the principal theorem. Namely the fact that the game is completely mixed implies that all optimal strategies are interior to the simplexes Q_1 and Q_2. Since there are no superfluous strategies there can be no interior bounding faces. This can only occur when the optimal strategies are unique.

SYMMETRIC GAMES

A symmetric game is one whose matrix A is skew-symmetric. Such games are of particular interest since these are the games in which, intuitively speaking, both players have the same rules or opportunities. Regarding these games Kaplansky [4] has observed that for such a game to be completely mixed it is necessary that the number of pure strategies be odd. The proof depends on the fact that the rank of a skew-symmetric matrix is even. By means of the principal theorem of this paper H. Kuhn has recently given a generalization of Kaplansky's theorem.

> THEOREM 3. Necessary and sufficient conditions that X and Y be solutions of a symmetric game are that, besides satisfying the conditions of the principal theorem,
>
> (a) the sets X and Y are isomorphic (that is, the transformation defined by $u_i \longrightarrow v_i$ carries X onto Y),
>
> (b) $e_1 - d_1 = e_2 - d_2$ must be odd.

> COROLLARY. If A is skew-symmetric and has unique solutions then e_1 and e_2 are odd. (Hence Kaplansky's result as a special case.)

BIBLIOGRAPHY

[1] WEYL, H., "The Elementary Theory of Convex Polyhedra," this Study.

[2] BOHNENBLUST, H. F., KARLIN, S., SHAPLEY, L. S., "The Solutions of Discrete, Two-person Games," this Study.

[3] BIRKHOFF, G., "Lattice Theory," American Math. Soc. Colloquium Publications, pp. 71-72.

[4] KAPLANSKY, I., "A Contribution to von Neumann's Theory of Games," Annals of Mathematics, Vol. 46 (1945), pp. 474-479.

D. Gale

S. Sherman

Princeton University and
The Institute for Advanced Study

SOLUTIONS OF DISCRETE, TWO-PERSON GAMES *

H. F. Bohnenblust, S. Karlin, and L. S. Shapley[1]

INTRODUCTION

In this paper we propose to investigate the structure of solutions of discrete, zero-sum, two-person games. For a finite game-matrix it is well known that a solution (i.e., a pair of frequency distributions describing the optimal mixed strategies of the two players) always exists (see [2][2], Chapter III, Section 17). Moreover, the set of solutions is known to be a convex polyhedron, each of whose vertices corresponds to a submatrix with special properties [3].

In Part I of the present paper we prove a fundamental relationship between the dimensions of the sets of optimal strategies, and devote particular attention to the set of games whose solutions are unique. Part II solves the problem of constructing a game-matrix with a given solution. A number of examples and geometrical arguments are interspersed to illustrate the theory, and Part III describes the solutions of some matrices with special diagonal properties.

PART I: STRUCTURE OF SOLUTIONS

§1. INTRODUCTION AND DEFINITIONS

Let Γ be the game described by the matrix $A = (a_{ij})$, with rows a_i and columns $a_j (i = 1, 2, \dots, m;\ j = 1, 2, \dots, n)$. Let X be the set of all optimal mixed strategies $x = (x_i)$ $(x_i \geq 0,\ 1 \cdot x = 1,\ \min_y xAy = v)$ of the maximizing player; Y that of the other player. Let $I_1(x)$ be the set of indices i for which $x_i > 0$, and $I_2(y)$ those for which $a_i \cdot y = v$; and similarly $J_1(y)$, $J_2(x)$. Then define

$$I_1 = \sum_{x \in X} I_1(x), \quad J_1 = \sum_{y \in Y} J_1(y)$$

$$I_2 = \prod_{y \in Y} I_2(y), \quad J_2 = \prod_{x \in X} J_2(x)$$

[1] Received January 24, 1949 by the ANNALS OF MATHEMATICS and accepted for publication; transferred by mutual consent to ANNALS OF MATHEMATICS STUDY No. 24.

[2] Numbers in square brackets refer to the bibliography at the end of the paper.

X_l = the set of x with $I_1(x) \subseteq I_l$, $l = 1, 2$;

Y_l = the set of y with $J_1(y) \subseteq J_l$, $l = 1, 2$.

Thus X_1 is the smallest face of the fundamental simplex of mixed strategies containing X, etc.

The purpose of Part I is to prove the relations:

(1) $$I_1 = I_2, \quad J_1 = J_2$$ (Theorem 1)

(2) $$\dim X_1 - \dim X = \dim Y_1 - \dim Y$$ (Theorem 2)

for all games with finite sets of pure strategies. We also show that the set of $m \times n$ game-matrices with unique solutions is dense and open in mn-space (Theorem 3). Under (2) we may subsume the corollary that a unique solution must be "square" — that is, involve the same number of pure strategies on each side. This is of especial interest since, by Theorem 3, games "in general position" have unique solutions.

For infinite matrices these results are not valid, even when I_1 and J_1 are finite. Nor does the analogue of Theorem 1 hold if the matrix is replaced by a continuous function, even though the pure strategies form compact sets. Simple examples supporting these assertions will be found in §7.

§2. REDUCTION TO THE ESSENTIAL PART OF THE GAME

LEMMA 1. There exists x in X such that $I_1(x) = I_1$ and $J_2(x) = J_2$ [y in Y such that $J_1(y) = J_1$ and $I_2(y) = I_2$].

PROOF. To each $i \in I_1$ and $j \in J_2$ there corresponds an $x'^{(i)}$ with $x_i'^{(i)} > 0$ and an $x''^{(j)}$ with $a_j \cdot x''^{(j)} = v$. The center of gravity of these $x'^{(i)}$ and $x''^{(j)}$ can be taken to be the x of the lemma.

LEMMA 2. $I_1 \subseteq I_2$ [$J_1 \subseteq J_2$].

PROOF. Choose x and y as in the preceding lemma. Then $\sum x_i(a_i \cdot y) = v$ implies that $x_i > 0$ only if $a_i \cdot y = v$. Hence $I_1 = I_1(x) \subseteq I_2(y) = I_2$.

Let Γ' be the game deduced from Γ by taking only the indices of I_2 and J_2. Any solution of Γ is a solution of Γ'; hence $v(\Gamma) = v(\Gamma')$ and

(3) $$X(\Gamma) \subseteq X(\Gamma'), \quad Y(\Gamma) \subseteq Y(\Gamma') .$$

LEMMA 3. With Γ' derived from Γ as above,

(4) $$\dim X(\Gamma) = \dim X(\Gamma'), \ \dim Y(\Gamma) = \dim Y(\Gamma') .$$

PROOF. Pick $x \in X(\Gamma)$ by Lemma 1 and pick any $x' \in X(\Gamma')$. Then there exists an x'' interior to the segment xx' which is in $X(\Gamma)$!

As simple consequences of (3) and (4), we observe that

(5) $$I_l, J_l, X_l, Y_l(\Gamma) = I_l, J_l, X_l, Y_l(\Gamma'), \quad l = 1, 2.$$

In view of (4) and (5), we need to prove the assertions (1) and (2) only for the smaller game Γ'. Or, what is the same thing, we may henceforth assume for Γ itself the properties:

(6) $$\text{for every } x \in X, \quad a_j \cdot x = v, \quad j = 1, 2, \ldots, n ;$$

(7) $$\text{for every } y \in Y, \quad a_i \cdot y = v, \quad i = 1, 2, \ldots, m .$$

Under these assumptions we shall verify (in § 4) that each player has an optimal mixed strategy to which every pure strategy contributes positive weight (Theorem 1). In § 3 we describe the geometrical motivation for the algebraic argument.

§ 3. GEOMETRIC ANALOGUE

The game-matrix A may be plotted in n-space as the convex U of the m points a_i. U is then the image under the linear transformation represented by A, of the fundamental simplex of mixed strategies of the x-player.

Suppose for simplicity that $v = 0$, and let Q be the "positive quadrant" in n-space. Then U has no interior point in common with Q, but these two convex polyhedra touch in some non-empty set T, the image of X. Under the reduction assumption (6), T will be precisely the origin. The optimal mixed strategies for the y-player will correspond to the hyperplanes through the origin which separate Q and U.[3] Reduction assumption (7) means that every separating plane actually contains the entire set U.

The parameter α of § 4 has the effect of shrinking U about an interior point $\bar{a}$. Lemma 5 states that U may be so shrunk and still maintain contact with Q. This leads easily to Theorem 1, which states

---

[3]The proof that a game has a solution (the minmax theorem) may be reduced to the proof that two convex sets with no interior points in common can always be so separated.

that the contact point is the center of gravity of a set of positive masses $[x_i > 0]$ at the vertices $[i \in I_2]$ of U.

§ 4. STUDY OF THE REDUCED GAME

Let Γ fulfill conditions (6) and (7). Put

$$\bar{a} = \frac{1}{m} \sum a_i ,$$

then for any $y \in Y$, $\bar{a} \cdot y = v$. Form the new game $\Gamma_\alpha = (b_{ij})$ with the rows

$$b_i = (1 - \alpha) a_i + \alpha \bar{a}, \quad 0 < \alpha < 1 .$$

For any $y \in Y$, $b_i \cdot y = v$ for each i, hence $v(\Gamma_\alpha) \leq v$.

LEMMA 4. If $Y_1(\Gamma_\alpha)$ and Y have a common point, then $v(\Gamma_\alpha) = v$.

PROOF. Take any y in both $Y_1(\Gamma_\alpha)$ and Y, and choose $y' \in Y(\Gamma_\alpha)$ by Lemma 1. The segment yy' may be extended beyond y' to $y'' = (1 + \varepsilon) y' - \varepsilon y$, $\varepsilon > 0$. Then

$$v(\Gamma_\alpha) \leq \max_i b_i \cdot y'' = (1 + \varepsilon)\, v(\Gamma_\alpha) - \varepsilon v$$

and thus $v \leq v(\Gamma_\alpha)$. But in any case $v(\Gamma_\alpha) \leq v$!

LEMMA 5. There exists $\alpha > 0$ such that $v(\Gamma_\alpha) = v$.

PROOF. By compactness there must exist a sequence $\{y^{(n)}\}$, $y^{(n)} \in Y(\Gamma_{\alpha_n})$, $\alpha_n \longrightarrow 0$, which converges to some mixed strategy y. Since the number of possible sets $Y_1(\Gamma_{\alpha_n})$ is finite, we may further stipulate that they all be the same (closed) set and thus all contain y. But because $b_i \cdot y^{(n)} \leq v$ for each i, n, we find that y is in Y. By Lemma 4 the desired α therefore exists.

THEOREM 1. $I_1 = I_2$ $[J_1 = J_2]$.

PROOF. Choose α by Lemma 5 and $x \in X(\Gamma_\alpha)$. Then the mixed strategy x', all of whose components

$$x_i' = (1 - \alpha) x_i + \frac{\alpha}{m}$$

are positive, satisfies

$$a_j \cdot x' = b_j \cdot x \geq v, \quad \text{all} \quad j ,$$

and is therefore in X. Hence I_1 contains every index i in the reduced game.

§ 5. THE FUNDAMENTAL THEOREM

THEOREM 2. $\dim Y_1 - \dim Y = \dim X_1 - \dim X$.

PROOF. We may suppose $v = 0$ without impairing the final result. The conditions defining X may be set forth:

(a) $a_j \cdot x = 0$ for all j ,

(b) $1 \cdot x = 1$,

(c) $x_i \geq 0$ for all i .

(a) and (b) together define a set C containing X; (a) alone defines a larger set C'. C' is in fact the null space of A, and therefore

(8) $$\dim C' = m - \text{rank } A.$$

Since the origin is in C' but not C, condition (b) actually lowers the dimension, i.e.,

(9) $$\dim C = \dim C' - 1 .$$

By Theorem 1 the inequalities (c) hold strictly for some x in X (and thus for a neighborhood); hence

(10) $$\dim X = \dim C .$$

Finally, and obviously,

(11) $$\dim X_1 = m - 1 .$$

Therefore, by (8) - (11):

(12) $$\text{rank } A = \dim X_1 - \dim X .$$

This, with the symmetrical expression for A^T, suffices to prove the theorem.

We observe for future use that in games whose value is not zero, the analogue of (12) is

(12) $$\text{rank } A = \dim X_1 - \dim X + 1 .$$

A, in these expressions, is of course the essential part (in the sense of § 2) of the total game-matrix.

The theorem may be interpreted also as a relationship between the zeros and the range of an operator and its adjoint. Let T and T^* denote the operator corresponding to A and its transpose, and let η, η^* and R, R^* denote the manifold of zeros and the range of T, T^*.

Then it is easily shown that

$$(13)\qquad \begin{aligned} \dim \eta^* &= \dim X \\ \dim \eta &= \dim Y . \end{aligned}$$

Now η^* is isomorphic to $R^\perp$, the orthogonal manifold to R; thus

$$\dim \eta^* = \dim X_1 - \dim R .$$

But $\dim R = \dim Y_1 - \dim \eta$, hence Theorem 2 is confirmed.

This point of view will be helpful in explaining the method of Part II.

§6. UNIQUE SOLUTIONS

Let U be the set of $m \times n$ game-matrices which have unique solutions (m and n are fixed throughout the discussion). Formally $A \in U$ if and only if

$$(14)\qquad \dim X(A) = \dim Y(A) = 0 .$$

We shall prove

THEOREM 3. U is open and everywhere dense in mn-space.

PROOF. (I) <u>Proof that</u> U <u>is open</u>.
If $\zeta = (\zeta_i)$ is a matrix or a vector, define

$$|\zeta| = \max_i |\zeta_i| .$$

LEMMA 6. v, X and Y are continuous functions of A in the following sense: Given A and any $\delta > 0$, there exists ε such that $|B - A| < \varepsilon$ implies

(a) $|v(B) - v(A)| < \delta$.

(b) $\displaystyle\min_{\substack{x \in X(A) \\ x' \in X(B)}} |x' - x| < \delta$.

(c) $\displaystyle\min_{\substack{y \in Y(A) \\ y' \in Y(B)}} |y' - y| < \delta$.

PROOF. Any $\varepsilon < \delta$ is suitable for (a). For (b) and (c), take any sequence $\{B^\mu\} \to A$ and choose for each μ some $x^\mu \in X(B^\mu)$. By compactness, every x^μ after some x^{μ_0} will be within δ of some accumulation point of $\{x^\mu\}$. But since

$$b_j^\mu \cdot x^\mu \geq v(B^\mu) \quad , \quad \text{all} \quad j, \mu ,$$

every accumulation point is in $X(A)$. An argument by contradiction now shows that the desired $\varepsilon > 0$ exists.

We now prove that U is open. Let A_1 be the essential part of A:

$$A_1 = (a_{ij}), \; i \in I_1 = I_2, \; j \in J_1 = J_2 \; .$$

Take $A \in U$. If $|B - A|$ is small, B has a solution x', y' near the unique solution x, y of A, by Lemma 6. (Two-sided uniqueness here makes an essential appearance in the proof. Thus the set U_X of games with unique optimal strategy for the first player is not open.) The components positive in x will be positive in x', hence

$$I_1(A) \subseteq I_1(B) \; .$$

Also Ay will be near By', hence

$$I_2(B) \subseteq I_2(A) \; .$$

It follows that A_1 and B_1 correspond and that

(15) $$X_1(A) = X_1(B) \; .$$

Now for any matrix C_0,

$$\lim_{C \to C_0} \text{rank } C \geq \text{rank } C_0 \; ;$$

we may therefore write

(16) $$\text{rank } B_1 \geq \text{rank } A_1 \; .$$

Assume $v(A) \neq 0$, then by (12, 12') of Part I,

(17) $$\begin{aligned} \dim X_1(A) - \dim X(A) &= \text{rank } A_1 - 1 \; , \\ \dim X_1(B) - \dim X(B) &\geq \text{rank } B_1 - 1 \; . \end{aligned}$$

Equations (15), (16), (17) give us

$$\dim X(B) \leq \dim X(A) \; .$$

After the same argument on Y, we conclude $B \in U$. Since the restriction $v \neq 0$ cannot be relevant in this context, every $A \in U$ has a neighborhood in U.

(A parallel proof could be given in operator terminology. Thus, the companion to (16) is the fact that the dimension of the zeros of an operator can only decrease for small perturbations.)

(II) <u>Proof that U is everywhere dense</u>.

We shall call a matrix A "general" if no $r \times r + 1$ submatrix of A or A^T, with a row 1 subjoined, has a vanishing determinant. The set G of general $m \times n$ matrices is evidently not empty for $mn > 1$.

LEMMA 7. G is everywhere dense.

PROOF. Take any A; and $C_\varepsilon = A + \varepsilon B$, $B \in G$. It suffices to find ε_0 such that $C_\varepsilon \in G$ for all positive $\varepsilon < \varepsilon_0$. Every determinant obtained from C_ε as in the preceding paragraph is a polynomial of rth degree in ε. Take $\varepsilon_0 > 0$ smaller than any positive root of these polynomials.

LEMMA 8. $G \subseteq U$.

PROOF. Take any $A \notin U$ with $y, y' \in Y$, $y \neq y'$; and let B be the submatrix A_1 with a row 1 subjoined. Then for any $x \in X$,

$$(18) \qquad (x, -v)^T B = 0^T .$$

Also

$$(19) \qquad B(y - y') = 0 .$$

Every submatrix of B having as many rows as B (by (18)), or as many columns (by (19)), must be singular. Therefore at least one of the determinants obtained from A as above vanishes, and A is not general.

The proof of Theorem 3 is completed by direct application of Lemmas 7 and 8.

§7. SOME GAMES WITH INFINITE SETS OF STRATEGIES

Virtually none of the foregoing theory applies without drastic modification to games with infinite payoff matrices. We submit here some examples[4] to justify this assertion. We shall not, however, enter into a systematic study at this time either of infinite game-matrices or of games with continua of strategies.

First we must observe that many infinite matrices do not possess values or optimal strategies, even approximately, and hence should perhaps

[4]For helpful comment and suggestions concerning these examples the authors are indebted to H. Kuhn of Princeton University.

not be called games at all. As a case in point, consider the unbounded matrix

$$A: \quad a_{ij} = i - j \quad (i = 1, 2, \ldots, \quad j = 1, 2, \ldots) .$$

This game has no value and hence no solution. For consider the mixed strategy x in which x_i is $1/i$ if $i = 2, 4, 8, \ldots,$ and zero otherwise. Then $a_j \cdot x = \infty$. By symmetry we are led to the curious result:

$$\inf_y \sup_i a_i \cdot y - \sup_x \inf_j a_j \cdot x = -\infty .$$

(For bounded matrices the corresponding expression is always non-negative. But with $|A| = \infty$, multiplication of matrices is no longer associative: $x^T(Ay) \neq (x^T A)y$.)

The matrix

$$B: \quad b_{ij} = a_{ij} / \sqrt{1 + a_{ij}^2} \qquad (a_{ij} \text{ as above}) ,$$

is bounded, but likewise has no value. For, given any y and $\varepsilon > 0$, we may choose n so that $y_1 + \ldots + y_n \geq 1 - \varepsilon$. Then for any $i > n/\varepsilon$, we have $1 \geq b_i \cdot y > 1 - 3\varepsilon$. By symmetry:

$$\inf_y \sup_i b_i \cdot y - \sup_x \inf_j b_j \cdot x = 2 .$$

Confining ourselves now to infinite game-matrices which do have solutions, we still find violations of all of our chief theorems. The following two games each have unique solutions.

$$C: \quad \begin{array}{llllll} 0 & c_1 - 1 & c_2 & c_3 & c_4 & \cdots \\ 0 & c_1 & c_2 - 1 & c_3 & c_4 & \\ 0 & c_1 & c_2 & c_3 - 1 & c_4 & \\ 0 & c_1 & c_2 & c_3 & c_4 - 1 & \\ \vdots & & & & & \ddots \end{array} \qquad c_k \geq c_{k+1} > 0 , \quad \Sigma c_k = 1 .$$

$$v(C) = 0; \; X(C) = \{(c_1, c_2, \ldots)\}; \; Y(C) = \{(1, 0, 0, \ldots)\} .$$

(To prove the last, we observe that any optimal y must play all strategies after the first with equal frequency — hence with zero frequency.)

$$D: \quad \begin{array}{llllll} d & 2d & 1/2 & 2d & 1/4 & \cdots \\ d & 1 & 2d & 1/3 & 2d & \cdots \end{array} \qquad d \neq 0 .$$

$$v(D) = d; \; X(D) = \{(1/2, 1/2)\}; \; Y(D) = \{(1, 0, 0, \ldots)\} .$$

Theorem 1 fails in the former game, since $J_2(C)$ is infinite, $J_1(C)$ finite. The dimensionality relation of Theorem 2 leads to $\infty = 0$ and $1 = 0$ in the two cases. Thus it breaks down even when the essential part of the game is a finite submatrix.

The continuity of solutions, as set forth in Lemma 6, is violated in the neighborhoods of both games. For example, given any $\varepsilon > 0$ we can move Y(D) a distance of 1 by subtracting ε from columns n and n + 1 of D, where $n > 1/\varepsilon$, $n \neq 1$.

The failure of Theorem 3 is illustrated again by both games. It is easy to verify that all x within ε of X(C) or within $\varepsilon/2d$ of X(D) become optimal if any $\varepsilon > 0$ is subtracted from the first column of C or D. Thus U is not open for infinite matrices. U continues to be everywhere dense in the subset of matrices which have solutions.

PART II: CONSTRUCTION OF GAMES WITH GIVEN SOLUTIONS

§8. THE PROBLEM: CANONICAL FORMS

We suppose that we have been given a pair of (convex) polyhedra X and Y, each situated in a simplex, and that we wish to find a game whose sets of optimal strategies correspond exactly to X and Y.

The problem is made no more difficult by prescribing the value the game is to have.

Let X_1 and Y_1 be the smallest faces of the simplices containing X and Y. From Theorem 2 we know that our problem has no answer unless

$$\dim X_1 - \dim X = \dim Y_1 - \dim Y . \tag{20}$$

Our subsequent work will show that condition (20) is sufficient as well (Theorem 4, §13).

The construction we describe in §10 - §12 produces a specific (except for a certain freedom in ordering the rows and columns) standardized matrix for each X, Y satisfying (20). This gives us in effect a set of canonical forms. However, there is no apparent way of relating an arbitrary matrix to its canonical counterpart short of finding all its solutions and constructing the standard matrix to order. (A finite though tedious process for finding all solutions is described in [3], §6.) These canonical forms, therefore, are not promising as a computational aid. In answering theoretical questions, however, it may sometimes be helpful to have to consider only a small subset of all possible matrices.

A more general classification might lump together games whose sets X, Y are isomorphic under a projective transformation. There would then be only a finite number of types for matrices of each particular size. The canonical games could have their solutions oriented in some natural, simple fashion; for example, the canonical unique solutions could always be of the form:

$$X = x^* = (1/r, \ldots, 1/r, 0, \ldots, 0) ,$$

$$Y = y^* = (1/r, \ldots, 1/r, 0, \ldots, 0) .$$

The solutions of any game would be identical to the solutions of a game BAC where A is one of a finite set of canonical games and B and C are nonsingular matrices representing the appropriate projective transformations on the two simplices of mixed strategies.

§9. GEOMETRICAL DESCRIPTION OF THE CONSTRUCTION

We return to the operator point of view introduced at the end of §5. Suppose first that the polyhedra given, X and Y, are in fact the intersections of the linear spaces $\overline{X}$ and $\overline{Y}$ with the fundamental simplices ((m - 1)- and (n - 1)-dimensional, respectively) lying in m- and n-dimensional Euclidean space. Suppose further that X and Y contain points interior to their respective simplices. The essential part of any game (see §2) will have such solutions.

In view of (20), the orthogonal manifolds $\overline{X}^{\perp}$ and $\overline{Y}^{\perp}$ have the same dimension. Take any (non-singular) linear transformation S mapping $\overline{X}^{\perp}$ on $\overline{Y}^{\perp}$, and any projection P of m-space on $\overline{X}^{\perp}$ which maps $\overline{X}$ into the origin. Then the game-matrix corresponding to the transformation T = SP has value zero, and sets of optimal strategies X and Y. (See §5, esp. equations (13).)

In general, the given polyhedra X and Y may have both "natural" faces, caused by the boundary inequalities $x_i \geq 0$, $y_j \geq 0$ of their simplices, and "unnatural" faces defined by arbitrary inequalities. Each unnatural face of X [Y] corresponds to a column [row] outside of the essential part of the game matrix. (See §12 below.)

Thus our ability to construct a game with given solutions is conditioned not only by the dimensional restriction (20) but also by whether we are provided with enough "dummy" strategies, not involved in any optimal strategy, to take care of the unnatural faces. It is always possible, of course, to handle a surplus of dummy strategies, without disturbing the rest of the construction.

§10. UNIQUE SOLUTION GIVEN

We proceed now to the algebraic details of the construction. In this section we suppose that the given X and Y are points. The general treatment of §11 includes this case: we consider it separately only to illustrate the general attack and because of the particular interest of games with unique solutions.

Disregard any dummy strategies, and denote the positive components of the unique optimal strategies by

$$x_0, x_1, \ldots, x_t = x, \quad t = \dim X_1 \ ;$$

$$y_0, y_1, \ldots, y_t = y, \quad\quad = \dim Y_1 \ .$$

The order may be taken arbitrarily (see Remark 3 below).

Let I^t represent the $t \times t$ identity matrix (1 on the main diagonal, zero elsewhere), and let $A(c)$ be the $t + 1 \times t + 1$ matrix as follows:

$$A(c) = \begin{pmatrix} a(c) & \frac{c-x_1}{x_0} & \frac{c-x_2}{x_0} & \cdots & \frac{c-x_t}{x_0} \\ \frac{c-y_1}{y_0} & & & & \\ \frac{c-y_2}{y_0} & & & & \\ \vdots & & I^t & & \\ \frac{c-y_t}{y_0} & & & & \end{pmatrix} \qquad a(c) = \frac{c(x_0+y_0-1) + \sum_1^t x_i y_i}{x_0 y_0}$$

Then $v(A(c)) = c$ and, provided $c \neq 1/t$, x, y is the unique solution of $A(c)$.

PROOF. One may verify directly that x, y is a solution and that c is the value. To establish uniqueness, we observe that

$$\operatorname{rank} A(c) = \begin{cases} t + 1, & c \neq 0, 1/t \ ; \\ t, & c = 0 \ . \end{cases}$$

Then by (12, 12') of § 5,

$$\dim X(A(c)) = \dim Y(A(c)) = 0 \ .$$

REMARK 1. In the event $c = 1/t$ the mixed strategies $x' = y' = (0, 1/t, \ldots, 1/t)$ are optimal as well as x and y. $X(A(1/t))$ and $Y(A(1/t))$ are the segments $x'x$ and $y'y$ extended to the edges of the simplices X_1 and Y_1. For a game with unique solution x, y and value $1/t$ we might use (for example) the matrix $B = A(2/t)/2$.

REMARK 2. A construction using any non-singular $t \times t$ matrix in place of I^t is equally possible. The generalization of the critical value $1/t$ is $1/\sum\sum b_{ij}$ where (b_{ij}) is the inverse of the matrix used.

REMARK 3. Since the numbering of the rows and columns is purely a matter of nomenclature, the matrix $A(c)$ is not wholly specific. We may remedy this defect by positing

$$x_0 \geq x_1 \geq \cdots \geq x_t, \quad y_0 \geq y_1 \geq \cdots \geq y_t .$$

§11. POLYHEDRA OF GENERAL DIMENSION GIVEN

We shall require in this section only that X [and Y] be completely described in some $(r - 1)$-dimensional [$(s - 1)$-dimensional] plane by the characteristic inequalities of the simplex:

$$x_i \geq 0, \quad [y_j \geq 0] .$$

When a game A has such sets of solutions, its essential part A_1 will have precisely the same sets, i.e.:

$$X(A) = X(A_1), \quad Y(A) = Y(A_1)$$

(compare (3) of §2). It will be most economical, then, to construct a game which is its own essential part. X_1 and Y_1 will comprise the full simplices of mixed strategies.

Define

$$r = 1 + \dim X, \quad m = 1 + \dim X_1 ,$$

$$s = 1 + \dim Y, \quad n = 1 + \dim Y_1 ,$$

$$t = m - r = n - s .$$

Choose r and s linearly independent points in X and Y respectively

$$x_k = (x_{k1}, \ldots, s_{km}) \quad k = 1, \ldots, r ,$$

$$y_l = (y_{l1}, \ldots, y_{ln}) \quad l = 1, \ldots, s ;$$

and use them as rows in forming the matrices

$$X^r_{rm} = (x_{ki}) \quad \text{and} \quad Y^s_{sn} = (y_{lj}) .$$

The superscripts show rank; the double subscripts show size. Our present problem may now be stated algebraically: to find A^{ρ}_{mn} satisfying

$$(21)\qquad \begin{cases} X^{r}_{rm} \; A^{\rho}_{mn} = v1_{rn} \\ A^{\rho}_{mn}(Y^{s}_{sn})^{T} = v1_{ms} \end{cases}$$

with

$$(22)\qquad \rho = \begin{cases} t, & v = 0 \;, \\ t + 1, & v \neq 0 \;. \end{cases}$$

The condition on ρ prevents $X(A)$, $Y(A)$ from being of higher dimension than the given X, Y (see (12, 12') in §5).

System (21) comprises $rn + ms$ equations in the mn unknowns a_{ij}. However, as the following construction reveals, there is just enough interdependence to permit t^2 of the unknowns to be chosen arbitrarily.

Before proceeding with the construction, it will be convenient to rearrange the columns of X^{r}_{rm} [and Y^{s}_{sn}] so that the first r [s] columns are linearly independent. Geometrically, this simply means picking the order of coordinates so that the projection of X on the simplicial face defined by the first r coordinates $(i = 1, \ldots, r)$ will be one-one.

Now we place I^t in the lower right corner of A (see the figure). The remaining elements a_{ij} are then determined uniquely by (21). Specifically,

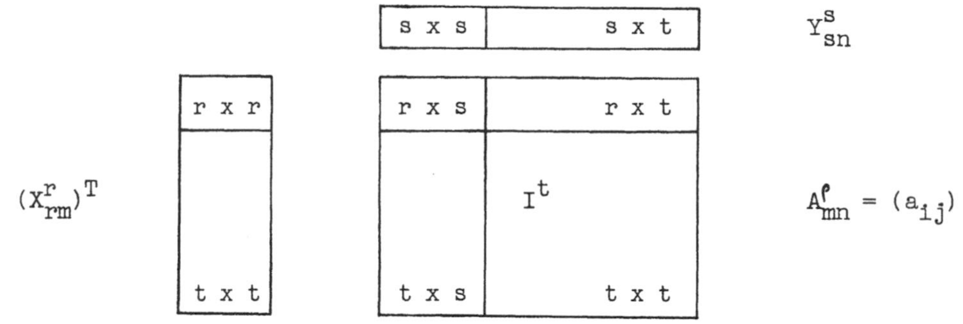

for the upper right corner $(i = 1, \ldots, r;\; j = s + 1, \ldots, n)$ use the systems

$$(23)\qquad \begin{cases} x_{11}a_{1j} + \cdots + x_{1r}a_{rj} = v - x_{1j} \;, \\ \vdots \\ x_{r1}a_{1j} + \cdots + x_{rr}a_{rj} = v - x_{rj} \;. \end{cases}$$

The lower left corner is analogous. The upper left may be filled in from either direction: the result must be the same. The finished matrix A is happily independent of our choice of points x_k and y_l.

A simple check reveals that condition (22) on the rank of A is fulfilled except when $v = 1/t$. In that instance we find the extraneous optimal strategy $x' = (0, \ldots, 0, 1/t, \ldots, 1/t)$. The restriction we put on the first r columns of X^r_{rm} tells us that x' is not in X. Therefore, if a value of $1/t$ is desired, we must use some such device as proposed in Remark 1, §10.

Remark 2 of §10 applies with equal force to the present case. As to the point raised in Remark 3, it is not clear that any generalization of the ordering proposed there would necessarily make the first r columns of X^r_{rm} [s columns of Y^s_{sn}] independent, as required.

§12. THE MOST GENERAL CASE

We must now consider given polyhedra whose boundaries are not entirely described by the natural limits $x_i \geq 0$, $y_j \geq 0$. Each unnatural $(r - 2)$-face of X [$(s - 2)$-face of Y] corresponds to a new column a_j [row a_i] outside of the essential submatrix A_1. There is no restriction on the number or arrangement of these unnatural faces, provided of course that X and Y remain convex. Furthermore, there is no interaction between the new columns $j \notin J_1$ and the new rows $i \notin I_1$: elements a_{ij} common to both may be assigned arbitrary values.

It suffices to describe the calculation of a_j for a particular $(r - 2)$-face F of X. Choose a set of $r - 1$ independent points $x_2, \ldots, x_r$ on F and another point x_1 in the interior of X. Let the distance of the latter to the plane of F be $\lambda > 0$. Form the matrix $F^r_{rm} = (x_{ki})$. The only condition that a_j must satisfy is

$$F^r_{rm} a_j = (v + \mu, v, \ldots, v) , \quad \mu > 0 .$$

To get a definite result we take $a_{ij} = 0$ for $i > r$, and $\mu = \lambda$. The latter makes the result independent of the points chosen. The former is justified because, after our manipulation in §11, the first r columns of F^r_{rm} constitute a nonsingular matrix F^r_{rr}. Thus we have

$$(24) \qquad F^r_{rr} a_j = (v + \lambda, v, \ldots, v) ,$$

which determines a_j exactly. The similar expression for the y-player involves $v - \lambda$ in place of $v + \lambda$. It is only at this point that the antisymmetry in the roles of the two players shows up in the construction.

§ 13. SUMMARY

Sections § 11, § 12, taken with Theorem 2, § 5, comprise a constructive proof of the following:

THEOREM 4. Let X be a convex polyhedron of dimension $r - 1$ contained in an $(m - 1)$-dimensional face X_1, but in no smaller face, of an $(m' - 1)$-dimensional simplex. Let there be just μ $(r - 2)$-dimensional faces of X not contained in the boundary of X_1. Similarly for Y, s, Y_1, n, n', and ν. Then an $m' \times n'$ game-matrix A exists having sets of optimal strategies corresponding exactly to X and Y if and only if

$$m - r = n - s$$

and

$$m' \geq m + \nu, \quad n' \geq n + \mu .$$

The complete construction may be summed up (see figure):

$A = (a_{ij})$:

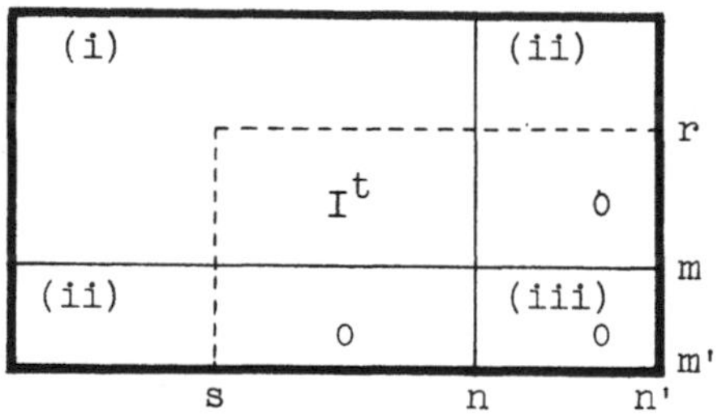

(i) Construct the essential submatrix A_1 around the identity matrix I^t, $t = m - r = n - s$, using equations (20) of § 11.

(ii) Compute each additional row or column required as outlined in § 12.

(iii) Square off the matrix by putting $a_{ij} = 0$, $i \notin I_1$, $j \notin J_1$.

§ 14. EXAMPLE

Find a game having value 2 and optimal strategies as indicated:

X:

$r = 2$
$m = 3$
$\mu = 1$

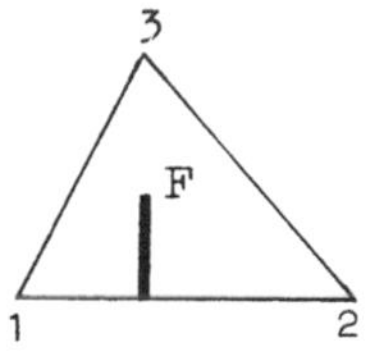

Extreme points (1/2, 1/2, 0)
(1/4, 1/4, 1/2)

In attempting to form X^r_{rm} we find we must reorder the coordinates: 1' = 2, 2' = 3, 3' = 1. Then

$$X^r_{rm} = \begin{array}{|ccc|} \hline 1/2 & 0 & 1/2 \\ 1/3 & 1/3 & 1/3 \\ \hline \end{array}$$

Y:

$s = 3$
$n = 4$
$\nu = 0$

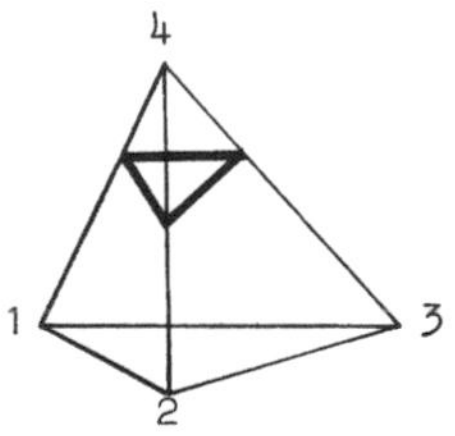

Extreme points (1/3, 0 , 0 , 2/3)
(0 , 1/2, 0 , 1/2)
(0 , 0 , 1/3, 2/3)

$$Y^s_{sn} = \begin{array}{|cccc|} \hline 1/3 & 0 & 0 & 2/3 \\ 0 & 1/2 & 0 & 1/2 \\ 0 & 0 & 1/3 & 2/3 \\ \hline \end{array}$$

Putting I^t in the lower right corner means setting $a_{3'4} = 1$. Column 4 is found by using X^r_{rm} (as in (23), §11), and the first three columns may be computed similarly using $(Y^s_{sn})^T$. By taking points with zero components in composing the matrices X^r_{rm} and Y^s_{sn} we were able to make the equations extremely simple.

X has an unnatural face F; we must therefore include a dummy strategy, $j = 5$, for the y-player. To find column 5 we select the points (1/3, 1/3, 1/3) interior to X and (1/4, 1/2, 1/4) in F. The distance between them is $\lambda = \sqrt{6}/12$. Substituting the matrix

$$F^r_{rr} = \begin{array}{|cc|} \hline 1/3 & 1/3 \\ 1/4 & 1/2 \\ \hline \end{array}$$

into (24) gives us $a_{1'5}$ and $a_{2'5}$. Finally, we set $a_{3'5} = 0$.

The completed game-matrix:

i: \ j:	1	2	3	4	5
1' = 2	0	1	0	3	$4 + \sqrt{6}/2$
2' = 3	2	2	2	2	$2 - \sqrt{6}/4$
3' = 1	4	3	4	1	0

PART III: SOLUTIONS OF SOME SPECIAL GAMES

§15. COMPLETELY MIXED GAMES; BASIC SOLUTIONS

Part III will describe the solutions of three easily recognizable types of square matrices with special diagonal properties. The first two (§16, §17) generalize the "separation of diagonals" criterion used in [2] (Chapter IV, Section 18) for solving 2 x 2 matrices, while the third (§18) is a special case of the second.

We shall find it convenient to introduce the notions of basic solution and completely mixed game. A solution x, y of A is basic if x is a vertex of $X(A)$ and y a vertex of $Y(A)$. If $v(A)$ is not zero, a solution x, y is basic if and only if there exists a non-singular submatrix of A whose inverse $B = (b_{ji})$ satisfies

(25) $$x_i = \sum_j b_{ji} / \sum_{i,j} b_{ji}, \quad y_j = \sum_i b_{ji} / \sum_{i,j} b_{ji} .$$

The value turns out to be

$$v = 1 / \sum_{i,j} b_{ji} .$$

(See reference [3], of which this is the main theorem.) The submatrix in question will contain the one defined by $I_1(x)$ and $J_1(y)$, and be contained in the one defined by $I_2(y)$ and $J_2(x)$ (see the definition of §1).

A game is said to be completely mixed if all of its solutions involve every strategy of both players (reference [1]). It follows that a completely mixed game must have a unique solution and a square matrix, which is non-singular unless the value is zero. The solution may be obtained by inverting the matrix and using (25).

LEMMA 9. All four of the game-matrices $\pm A$, $\pm A^T$, or none of them, are completely mixed.

The proof is routine.

§16. MAIN DIAGONAL SEPARATED AND DOMINANT

Consider an $n \times n$ matrix $A = (a_{ij})$ which satisfies, for some fixed q,

(26) $$\begin{cases} a_{ij} > q & \text{if } i = j , \\ a_{ij} < q & \text{if } i \neq j ; \end{cases}$$

and either (a):

$$\sum_i a_{ij} \geq nq \quad \text{for all} \quad j \; ,$$

or (b):

$$\sum_j a_{ij} \geq nq \quad \text{for all} \quad i \; .$$

Then $v(A) \geq q$ and A is completely mixed. (See Fig, 1, in which q is taken to be zero.)

PROOF. (a) Putting $x_1 = \ldots = x_n = 1/n$ reveals that $v \geq q$. Should any y in Y have $y_l = 0$, then the inequality $\sum a_{lj}y_j < q \leq v$ tells us that $x_l = 0$ for every x in X. But whenever $x_l = 0$ the inequality $\sum a_{il}x_i < v$ prevents x from being optimal. Thus A is completely mixed.

PROOF. (b) Take any y in Y and l such that $y_l = \max y_j$. Then

$$v \geq \sum a_{lj}y_j \geq q \; .$$

The proof continues as in (a).

Cases (a) and (b) might have been deduced one from the other with the aid of Lemma 9. Similarly we may reverse the inequalities of the hypothesis. The following example shows that the condition "either (a) ... or (b) ...", which ensures "uniform" dominance, cannot be replaced by the weaker proviso:

(27)
$$\sum_{i,j} a_{ij} \geq n^2 q \; .$$

EXAMPLE 1.

1	-2	0
-2	1	0
0	0	12

$v = 0$

unique $x \in X$: $(0, 0, 1)$

vertices of Y: $(1/3, 2/3, 0)$, $(2/3, 1/3, 0)$.

The main diagonal is separated, and dominant in the sense of (27) for every q satisfying (26); yet the game is not completely mixed.

§ 17. DIAGONALS SEPARATED AND ORDERED

Consider an $n \times n$ matrix $A = (a_{ij})$ with

$$a_{ij} \in L_k, \; k \equiv j - i \pmod{n} \; ,$$

where the intervals L_k are disjoint and ordered:

(28)
$$L_0 < L_1 < \ldots < L_{n-1} \; .$$

We shall establish that A is completely mixed. (See Fig. 2.)

PROOF. Suppose some $x \in X$ has $x_l = 0$. Then

$$\sum_i a_{il}x_i > \sum_i a_{i,l-1} \geq v .$$

Hence every $y \in Y$ has $y_l = 0$. Similarly, if some $y \in Y$ has $y_l = 0$ then

$$\sum_j a_{l+1,j}y_j < \sum_j a_{lj}y_j \leq v .$$

Hence every x in X has $x_{l+1} = 0$. Repeat these two steps n times, reducing subscripts modulo n when necessary. The resulting absurdity $x = y = 0$ proves that A is completely mixed.

Without the ordering (28), A is not necessarily completely mixed. In any case, use of the mixed strategy $(1/n, \ldots, 1/n)$ cannot cost either player more than the mean diameter of the sets L_k. The next section solves the case where the L_k are points.

$$\begin{array}{|cccc|c} + & - & \cdots & - & \geq 0 \\ - & + & \cdots & - & \geq 0 \\ \vdots & & & & \vdots \\ - & - & \cdots & + & \geq 0 \\ \hline \geq 0 & \geq 0 & \cdots & \geq 0 & \end{array}$$

(a) (b)

Figure 1

$$\begin{array}{|cccc|} L_0 & L_1 & \cdots & L_{n-1} \\ L_{n-1} & L_0 & \cdots & L_{n-2} \\ \vdots & & & \\ L_1 & L_2 & \cdots & L_0 \end{array}$$

Figure 2

$$\begin{array}{|cccc|} a_0 & a_1 & \cdots & a_{n-1} \\ a_{n-1} & a_0 & \cdots & a_{n-2} \\ \vdots & & & \\ a_1 & a_2 & \cdots & a_0 \end{array}$$

Figure 3

§18. CONSTANT DIAGONALS

Consider an $n \times n$ matrix $A = (a_{ij})$ with

$$a_{ij} = a_k, \quad k \equiv j - i \pmod{n} .$$

(See Fig. 3.) One immediately observes that $(1/n, \ldots, 1/n)$, $(1/n, \ldots, 1/n)$ is a solution and that $v(A) = (\sum a_k)/n$. Since any other optimal mixed strategies for either player must be symmetrically disposed about $(1/n, \ldots, 1/n)$, this solution will be unique if and only if it is basic. Suppose $\sum a_k \neq 0$. Then A is completely mixed if and only if the determinant $|a_{ij}| \neq 0$. But it is easily verified that

$$|a_{ij}| = \prod_{l=0}^{n-1} \sum_{k=0}^{n-1} a_k \omega_n^{kl}$$

where ω_n is a primitive nth root of unity. If none of the factors vanishes, then A is completely mixed. On the other hand, if $\sum a_k \omega_n^{kl} = 0$

for $l = l_o$, the real part r_{l_o} of the complex vector

$$(1, \omega_n^{l_o} \omega_n^{2l_o}, \ldots, \omega_n^{(n-1)l_o})$$

is in the null-space of A and A^T. The optimal mixed strategies for either player will be just those of the form

$$(1/n, \ldots, 1/n) + r$$

where r is a vector in the space spanned by all such r_{l_o} and their cyclic permutations.

EXAMPLE 2.

0	1	3	2
2	0	1	3
3	2	0	1
1	3	2	0

$v = 3/2$, $r_{l_o} = r_2 = (1, -1, 1, -1)$.

X and Y are the line segments, in their respective tetrahedra, joining the points $(1/2, 0, 1/2, 0)$ and $(0, 1/2, 0, 1/2)$.

§19. CONCLUSION

REMARK 1. Any matrix which is derivable from one of the types here discussed by permutation of the rows or columns is of course not essentially different.

REMARK 2. A relaxing of the strict inequalities appearing in §16, §17 gives rise to a host of special cases, most of them not completely mixed, which are not worth describing in detail.

REMARK 3. A 2×2 game is completely mixed if and only if its diagonals are separated. Unfortunately, our generalized conditions of §16, §17 are not even broad enough to cover all 3×3 completely mixed games. One of the mavericks is the following:

EXAMPLE 3.

4	-3	-2
-3	4	-2
0	0	1

$v = 1/7$

unique $x \in X$: $\frac{1}{7}(1, 1, 5)$

unique $y \in Y$: $\frac{1}{7}(3, 3, 1)$

Only the main diagonal is separated and it is not dominant.

BIBLIOGRAPHY

[1] KAPLANSKY, I., "A Contribution to von Neumann's Theory of Games," Annals of Mathematics, Vol. 46 (1945), pp. 474-479.

[2] MORGENSTERN, O., von NEUMANN, J., "Theory of Games and Economic Behavior," 2nd ed. (1947), Princeton University Press, Princeton.

[3] SHAPLEY, L., SNOW, R.,"Basic Solutions of Discrete Games," this Study.

H. F. Bohnenblust

S. Karlin

L. S. Shapley

The RAND Corporation

SOLUTIONS OF GAMES BY DIFFERENTIAL EQUATIONS*

G. W. Brown and J. von Neumann

§1.

The purpose of this note is to give a new proof for the existence of a "value" and of "good strategies" for a zero-sum two-person game. This proof seems to have some interest because of two distinguishing traits:

(a) Although the theorem to be proved is of an algebraical nature, a very simple proof is obtained by analytical means.

(b) The proof is "constructive" in a sense that lends itself to utilization when actually computing the solutions of specific games. The procedure could be "mechanized" with relative ease, both for "digital" and for "analogy" methods. In the latter case it is probably much less sensitive to the precision of the equipment, than the somewhat related problem of "linear equation solving" or "matrix inversion."

The derivations which follow are based on results that were obtained independently by the two authors. Further results of one of them (G. W. Brown) are published elsewhere [1][1].

§2.

Consider first the special case of a symmetric game, i.e., where the "game" matrix" $A_{ij}(i, j = 1, \ldots, m)$ is antisymmetric: $A_{ij} = -A_{ji}$.

Write for vectors, $x = (x_i)$, $u = (u_i)$, and use the notations

$$(1)\qquad \begin{cases} u_i = \sum_j A_{ij}x_j , \\ \varphi(u_i) = \operatorname{Max}(0, u_i) , \\ \phi(x) = \sum_i \varphi(u_i) , \\ \Psi(x) = \sum_i (\varphi(u_i))^2 . \end{cases}$$

Consider the differential equation system

$$(2)\qquad \frac{dx_i}{dt} = \varphi(u_i) - \phi(x) \cdot x_i ,$$

[1] Numbers in square brackets refer to the bibliography at the end of this paper.

*Accepted as a direct contribution to ANNALS OF MATHEMATICS STUDY No. 24.

starting with a vector

(3) $$x^0 = (x_i^0),\ x_i^0 \geq 0,\ \sum_i x_i^0 = 1\ .$$[2]

$x_i = 0$ implies $\frac{dx_i}{dt} = \varphi(u_i) \geq 0$, hence $x_i \geq 0$ can never go over into $x_i < 0$, i.e., always

(4) $$x_i \geq 0\ .$$

Summing over all i gives

$$\frac{d}{dt}\left(\sum_i x_i\right) = \phi(x)\left(1 - \sum_i x_i\right)\ ,$$

i.e.,

$$\frac{d}{dt} \ln \left|1 - \sum_i x_i\right| = -\ \phi(x)\ ,$$

hence $\sum_i x_i = 1$ can never go over into $\sum_i x_i \neq 1$, i.e., always

(5) $$\sum_i x_i = 1\ .$$

Next, when $\varphi(u_i) > 0$, then

$$\frac{d\varphi(u_i)}{dt} = \sum_j A_{ij} \frac{dx_j}{dt} = \sum_j A_{ij}\ \varphi(u_j) - \phi(x) \sum_j A_{ij} x_j\ ,$$

hence always

$$\frac{d(\varphi(u_i))^2}{dt} = 2 \sum_j A_{ij}\ \varphi(u_i)\ \varphi(u_j) - 2\ \phi(x) \sum_j A_{ij}\ \varphi(u_i) x_j\ .$$

Summing over all i gives

$$\frac{d\Psi(x)}{dt} = 2 \sum_{ij} A_{ij}\ \varphi(u_i)\ \varphi(u_j) - 2\ \phi(x) \sum_{ij} A_{ij}\ \varphi(u_i) x_j\ .$$

The first term on the right-hand side vanishes, because A_{ij} is antisymmetric. The second term is equal to

$$\sum_i \varphi(u_i) \sum_j A_{ij} x_j = \sum_i \varphi(u_i) u_i\ .$$

---

[2]Existence of a unique solution to this system is assured by virtue of the fact that the system is piecewise well-behaved, with matching of first derivatives at the boundaries. The related system obtained by dropping the last term in (2) is piecewise a linear system, and has a growing solution, proportional to the solution of (2). The last term in (2) simply normalizes the solution to make (5) hold.

Whenever $\varphi(u_i) \neq 0$, then $\varphi(u_i) = u_i$. Hence the above expression is equal to $\sum_i (\dot{\varphi}(u_i))^2 = \Psi(x)$. Therefore

$$\frac{d\Psi(x)}{dt} = -2\,\phi(x)\,\Psi(x)\ . \tag{6}$$

Now clearly

$$(\Psi(x))^{\frac{1}{2}} \leq \phi(x) \leq (m\Psi(x))^{\frac{1}{2}}\ . \tag{7}$$

Hence as long as $\Psi(x) > 0$, also $\phi(x) > 0$, and $\Psi(x)$ is decreasing; also

$$\frac{d\Psi(x)}{dt} \leq -2\,(\Psi(x))^{\frac{3}{2}}, \quad -\frac{1}{2}\,(\Psi(x))^{-\frac{3}{2}}\,\frac{d\Psi(x)}{dt} \geq 1\ ,$$

hence

$$(\Psi(x))^{-\frac{1}{2}} \geq (\Psi(x_0))^{-\frac{1}{2}} + t\ ,$$

$$\Psi(x) \leq \frac{\Psi(x_0)}{\left(1 + (x_0)^{\frac{1}{2}}\,t\right)^2}\ . \tag{8}$$

If ever $\Psi(x) = 0$, then this remains true from then on (i.e., for all larger t), and so (8) is true again.

Finally from (7), (8)

$$\phi(x) \leq \frac{m^{\frac{1}{2}}\,\Psi(x_0)^{\frac{1}{2}}}{1 + \Psi(x_0)^{\frac{1}{2}}\,t} \tag{9}$$

and from (8)

$$\varphi(u_i) \leq \frac{\Psi(x_0)^{\frac{1}{2}}}{1 + \Psi(x_0)^{\frac{1}{2}}\,t}\ . \tag{10}$$

§ 3.

By (8), (9), (10) $t \longrightarrow +\infty$ implies

$$\Psi(x) \longrightarrow 0,\ \phi(x) \longrightarrow 0\ ,$$

and all

$$\varphi(u_i) \longrightarrow 0\ .$$

That the x_i themselves have limits for $t \to +\infty$ is not clear; (2) and (10) do not seem to suffice to prove this. Nevertheless, since the range (4), (5) of the x_i is compact, limit points $x^\infty = (x_i^\infty)$ of the $x = (x_i)$ for $t \to +\infty$ must exist. For any such x^∞ (4), (5) must again be true, and (10) gives that all

$$\varphi(u_i^\infty) = 0 ,$$

i.e., all

$$\sum_j A_{ij} x_j^\infty \leq 0 . \tag{11}$$

Hence any $x^\infty = (x_i^\infty)$ represents a "good strategy," and the "value" of the (symmetric) game is, as it should be, zero.

§4.

Consider next an arbitrary game, i.e., one with an unrestricted "game matrix" B_{kl} $(k = 1, \ldots, p;\ l = 1, \ldots, q)$. Various ways of reducing this to a symmetric game are known. They differ from each other, among other things, in the order m of the symmetric game, i.e., of the antisymmetric matrix $A_{ij}(i, j = 1, \ldots, m)$ to which they lead. A very elegant method of this type has been lately found [2] which gives the remarkably low value $m = p + q + 1$. One of the authors (J. von Neumann) had obtained earlier another method, which gives the larger value $m = pq$. (This is referred to, but not described, in [3], page 168.) We will follow here the second procedure, partly because its underlying qualitative idea is simpler, and partly because, although its m is considerably larger than that of the other method (pq vs. $p + q + 1$, cf. above), it leads ultimately to a set of differential equations in fewer variables ($p + q - 2$ vs. $p + q$, cf. the remarks at the end of §5.).

The qualitative idea behind the method referred to is this: Assume that a player knew how to play every conceivable symmetric game A. Assume that he were asked to play a (not necessarily symmetric) game B. How could he then reduce it to known (symmetric) patterns?

He could do it like this: He could imagine that he is playing (simultaneously) two games B: Say B' and B". In B' he has the role of the first player, in B" he has the role of the second player – for his opponent the positions are reversed. The total game A, consisting of B' and of B" together, is clearly symmetric. Hence the player will know how to play A – hence also its parts, say B', i.e., B.

In spite of the apparent "practical" futility of this "reduction," it nevertheless expresses a valid mathematical procedure. The mathematical procedure is as follows:

Let k', l' be the indices k, l for the game B', and k'', l'' those for the game B''. The player under consideration then controls the indices k', l'', and his opponent controls k'', l'. Hence i may be made to correspond to the pair (k', l''), and j to the pair (k'', l'). The game matrices are $B_{k'l'}$ for B' and $-B_{k''l''}$ for B'', i.e. $B_{k'l'} - B_{k''l''}$ for A, i.e.

(12) $A_{ij} = B_{k'l'} - B_{k''l''}$, where $i = (k', l'')$, $j = (k'', l')$.

The symmetry of this new game, i.e. the antisymmetry of A_{ij}, is obvious. Clearly $m = pq$.

Hence a system $x = (x_i) = (x_{kl})$ exists, such that all

$$x_i \geq 0, \sum_i x_i = 1 ,$$

and all

$$\sum_j A_{ij} x_j \leq 0$$

(cf. (4), (5), (11)). This means

$$x_{kl} \geq 0, \sum_{kl} x_{kl} = 1 ,$$

and

$$\sum_{k''l'} (B_{k'l'} - B_{k''l''}) \, x_{k''l'} \leq 0 ,$$

i.e.

$$\sum_{k''l'} B_{k'l'} x_{k''l'} \leq \sum_{k''l'} B_{k''l''} x_{k''l'} ,$$

$$\sum_{l'} B_{k'l'} \left(\sum_{k''} x_{k''l'}\right) \leq \sum_{k''} B_{k''l''} \left(\sum_{l'} x_{k''l'}\right) .$$

Putting

(13) $$\xi_k = \sum_l x_{kl}, \quad \eta_l = \sum_k x_{kl} ,$$

these inequalities yield

(14) $$\xi_k \geq 0, \quad \eta_l \geq 0 ,$$

(15) $$\sum_k \xi_k = 1, \quad \sum_l \eta_l = 1 ,$$

and

$$\sum_{l'} B_{k'l'} \eta_{l'} \leq \sum_{k''} B_{k''l''} \xi_{k''} ,$$

i.e.

(16) $$\operatorname{Max}_k \sum_l B_{kl}\eta_l \leq \operatorname{Min}_l \sum_k B_{kl}\xi_k .$$

(14), (15), (16) imply, of course, that equality holds in (16), that $\xi = (\xi_k)$ and $\eta = (\eta_l)$ are "good strategies" for the two players, and that the common value of both sides of (16) is the "value" of the (original, not necessarily symmetric) game. (Cf. [3], pp. 153 and 158.)

§ 5.

Apply now the differential equation system (2) to the "derived" game (12). Restating (1), (2) gives

$$u_{k'l''} = \sum_{k''l'} (B_{k'l'} - B_{k''l''})\, x_{k''l'} ,$$

$$\varphi(u) = \operatorname{Max}(0, u)$$

$$\phi(x) = \sum_{kl} \varphi(u_{kl})$$

$$\Psi(x) = \sum_{kl} (\varphi(u_{kl}))^2$$

and

$$\frac{dx_{kl}}{dt} = \varphi(u_{kl}) - \phi(x)\, x_{kl} .$$

This system involves $m = pq$ variables x_{kl}. It can, however, be "contracted" as follows:

Clearly

$$u_{k'l''} = \sum_{l'} B_{k'l'}\eta_{l'} - \sum_{k''} B_{k''l''}\xi_{k''} ,$$

i.e.

$$u_{kl} = v_k - w_l ,$$

where

(17) $$\begin{cases} v_k = \sum_l B_{kl}\eta_l , \\ w_l = \sum_k B_{kl}\xi_k . \end{cases}$$

$\varphi(u)$ may be defined as before:

(18) $$\varphi(u) = \operatorname{Max}(0, u) .$$

$\phi(x), \Psi(x)$ depend no longer on all pq components of $x = (x_i) = (x_{kl})$, but only on the $p + q$ components of $\xi = (\xi_k)$ and $\eta = (\eta_l)$:

(19)
$$\begin{cases} \Phi(\xi, \eta) = \sum_{kl} \varphi(v_k - w_l) , \\ \Psi(\xi, \eta) = \sum_{kl} (\varphi(v_k - w_l))^2 . \end{cases}$$

Summing the x_{kl}-differential-equations over all l gives

(20)
$$\frac{d\xi_k}{dt} = \sum_{l} \varphi(v_k - w_l) - \Phi(\xi, \eta)\, \xi_k ,$$

summing them over all k gives

(21)
$$\frac{d\eta_l}{dt} = \sum_{k} \varphi(v_k - w_l) - \Phi(\xi, \eta)\, \eta_l .$$

Thus a system (20), (21) has been obtained, which involves only $p + q$ variables ξ_k and η_l.

Combining the observations of § 3. and those at the end of § 4. shows, since the $\xi = (\xi_k)$, $\eta = (\eta_l)$ vary in the compact joint range defined by (14), (15), that they possess (joint) limiting points $\xi^\infty = (\xi_k^\infty)$, $\eta^\infty = (\eta_l^\infty)$. Any such pair represents a pair of "good strategies."

Because of (15), the number of variables involved in (2) is not m, but $m - 1$. Because of (15), the number of variables involved in (20), (21) is not $p + q$, but $p + q - 2$.

BIBLIOGRAPHY

[1] BROWN, G. W., Proceedings of Linear Programming Conference of the Cowles Commission (1950), The University of Chicago.

[2] GALE, D., KUHN, H., TUCKER, A. W., "On Symmetric Games," this Study.

[3] MORGENSTERN, O., von NEUMANN, J., "Theory of Games and Economic Behavior," 2nd ed. (1947), Princeton University Press, Princeton.

G. W. Brown
J. von Neumann

The RAND Corporation and
The Institute for Advanced Study

ON SYMMETRIC GAMES*

D. Gale, H. W. Kuhn and A. W. Tucker[1]

A symmetric game is a game with a skew-symmetric payoff matrix; informally, the two players play the same role in a symmetric game and have the same set of available pure strategies. Possessed of natural interest because of their special character, symmetric games are given additional importance by the computational procedures which are discussed by G. W. Brown and J. von Neumann in their contribution to this Study. In this short note we will investigate two methods for symmetrizing an arbitrary game; by this, we mean constructing a symmetric game whose solution yields a solution to the original game by a process which is trivial computationally.

We also describe an exceptionally concise proof of the fundamental theorem of the zero-sum, two-person game, which is completely algebraic and is provided by such symmetrizations. Finally, a characterization of the possible solution convexes for symmetric games is given.

§1. SYMMETRIZATION OF VON NEUMANN

Perhaps the simplest symmetrization has been suggested by von Neumann. If we are given the m by n game matrix $A = (a_{ij})$ we form the game described by the mn by mn matrix $S_1 = (a_{ij|kl})$ where $a_{ij|kl} = a_{il} - a_{kj}$ for $i, k = 1, \ldots, m$; $j, l = 1, \ldots, n$. Then $a_{ij|kl} = - a_{kl|ij}$ and hence S_1 is skew-symmetric. Let u be an optimal strategy for S_1, that is, a vector $u = (u_{ij})$ with $u_{ij} \geq 0$ and $\sum_{i,j} u_{ij} = 1$ such that: $\sum_{i,j} u_{ij} a_{ij|kl} \geq 0$.

Hence

$$\sum_{i,j} u_{ij} a_{ij|kl} = \sum_{i,j} u_{ij}(a_{il} - a_{kj})$$

$$= \sum_{i,j} u_{ij} a_{il} - \sum_{i,j} u_{ij} a_{kj} \geq 0$$

so that if we let $x_i = \sum_j u_{ij}$ and $y_j = \sum_i u_{ij}$ we have

(1) $$\sum_i x_i a_{il} \geq \sum_j a_{ij} y_j \ .$$

---

[1] Princeton and Stanford Universities, under contracts with the Office of Naval Research.

*Accepted as a direct contribution to ANNALS OF MATHEMATICS STUDY No. 24.

But $x = (x_i)$ and $y = (y_j)$ are mixed strategies for A and (1) implies

$$\min_j \Sigma_i x_i a_{ij} \geq \max_j \Sigma_j a_{ij} y_j \ .$$

Hence x and y are optimal strategies for A.

This symmetrization has a simple verbal description. Instead of playing the original game A, the players hold two plays of A simultaneously, each taking the role of player I in one play, of player II in the other. It is interesting to note that chess is usually symmetrized by a different device which yields exactly half of the matrix of S_1. Namely, as customarily played, chess is preceded by a chance move in which one player guesses the color of a concealed piece to decide which player plays first. Thus to play this game, both players must choose one strategy for each of the two equally probable outcomes of the chance move and the symmetrized matrix is:

$$S_2 = \tfrac{1}{2} S_1 \ .$$

§ 2. ANOTHER METHOD OF SYMMETRIZATION

If we are symmetrizing games for computational purposes, the size of the resulting matrix is clearly of the utmost importance. Although von Neumann has proposed a modification of one of his computational methods to reduce the number of steps required to solve the mn by mn game S_1 to essentially the number necessary for the solution of an $(m + n)$ by $(m + n)$ game, it is interesting that such a symmetrization can be made directly.

Here, given an m by n game matrix $A = (a_{ij})$, we form the symmetric game described by

$$S_3 = \begin{pmatrix} 0 & A & -1 \\ -A' & 0 & 1 \\ 1 & -1 & 0 \end{pmatrix} ,$$

where $-A'$ is the negative transpose of A, ± 1 are vectors composed of the appropriate number of ± 1's and the 0's represent matrices composed entirely of zeros filling out the remainder of S_3. This symmetrization was suggested by a result of G. W. Brown and G. B. Dantzig dealing with the equivalence of games and linear programming problems.

As is well known, we may arrange that v, the value of A, be positive by adding a positive constant without changing the sets of optimal strategies. Hence we assume, without restriction, that $v > 0$.

Now, if $(x_i;\ y_j;\ \lambda)$ is an optimal mixed strategy for S_3, we have

(2) $$-\sum_j a_{ij} y_j + \lambda \geq 0$$

(3) $$\sum_i x_i a_{ij} - \lambda \geq 0$$

(4) $$-\sum_i x_i + \sum_j y_j \geq 0 \quad .$$

Suppose $\sum_j y_j = 0$. Then, by (4), $\sum_i x_i = 0$ and (2) implies $\lambda = 0$ which contradicts $\sum_i x_i + \sum_j y_j + \lambda = 1$. Hence $\sum_j y_j = \mu > 0$ and $y^* = y/\mu$ is a mixed strategy for A.

Suppose $\lambda = 0$. Then (2) becomes $\sum_j a_{ij} y_j^* \leq 0$ contradicting our assumption that the value of A be positive. Hence $\lambda > 0$ and consequently multiplying each inequality (2) by x_i, each inequality (3) by y_j and summing, we obtain:

(5) $$\sum_i x_i \geq \frac{1}{\lambda} \sum_{i,j} x_i a_{ij} y_j \geq \sum_j y_j \quad .$$

Combining (4) and (5), $\sum_i x_i = \sum_j y_j = \mu$ and if we let $x^* = x/\mu$ inequalities (2) and (3) become:

(2^*) $$\sum_j a_{ij} y_j^* \leq \lambda/\mu$$

(3^*) $$\sum_i x_i^* a_{ij} \geq \lambda/\mu \, .$$

Thus, x^* and y^* are optimal for A and $v = \lambda/\mu$.

To give a verbal description of this symmetrization we consider a game in which the players are denoted by white and black. We assume that white has an advantage (i.e., if white is the first player, $v > 0$). Then the symmetrized game is given by the following rules:

The players choose independently to play white or black, or to hedge. If they choose the same colors or both hedge the play is a draw. If they choose different colors a play of the original game ensues. As for the remaining possibilities, a hedge wins one unit from white and loses one unit to black.

It is evident that this is a symmetric game. That we learn how to play the original game by playing the symmetrized version follows from the fact that an optional strategy for the symmetric game must include playing both white and black with positive probability. The cyclic nature of the possibilities (white, black, and hedge), reminiscent of Stone, Scissors, and Paper, makes this intuitively plausible.

§3. ALGEBRAIC PROOF OF MAIN THEOREM

A little known result of Erich Stiemke [see item 1 in bibliography appended] combined with either of the above symmetrizations provides a direct proof of the fundamental theorem of the zero-sum, two-person game; since Stiemke's proof is algebraic, the resulting proof is also. Indeed, since Stiemke places no more restrictions on the elements of the matrices involved than that they lie in an <u>ordered field</u>, the proof reveals that the value of the game and the components of the basic optimal strategies also lie in this field.

STIEMKE'S THEOREM.[2] Let $A = (a_{ij})$ be an m by n matrix. Then, either $\sum_i u_i a_{ij} \geq 0$ for some vector $u = (u_i)$ or $\sum_j a_{ij} v_j = 0$ for some vector $v = (v_j) > 0$.

If we apply Stiemke's theorem to the matrix $A = (S, I)$ where $S = (s_{ij})$ is <u>any</u> m by n matrix and I is the m by n unit matrix we have:

COROLLARY 1. For an arbitrary matrix $S = (s_{ij})$, either $\sum_i u_i s_{ij} \geq 0$ for some vector $u = (u_i) \geq 0$ or $\sum_j s_{ij} v_j < 0$ for some vector $v = (v_j) > 0$.

But if we restrict S to be <u>skew-symmetric</u>, $m = n$ and $\sum_j s_{ij} v_j < 0$ is equivalent to $\sum_i v_i s_{ij} > 0$. Hence we obtain:

COROLLARY 2. For a skew-symmetric matrix $S = (s_{ij})$, either $\sum_i u_i s_{ij} \geq 0$ for some vector $u = (u_i) \geq 0$ or $\sum_i v_i s_{ij} > 0$ for some vector $v = (v_i) > 0$.

Thus the first alternative always holds and if we normalize the vector u so that $\sum_i u_i = 1$, we have:

COROLLARY 3. For a skew-symmetric matrix $S = (s_{ij})$, there exists a vector $u = (u_i) \geq 0$ with $\sum_i u_i = 1$ such that $\sum_i u_i s_{ij} \geqq 0$.

But this is precisely the statement of the existence of an optimal strategy for the first player in a symmetric game, which was our point of departure in the first symmetrization, and hence proves the fundamental theorem for an abritrary game.

[It should be remarked that Stiemke's Theorem and Corollaries 1-3 represent the same pattern found in "The Theory of Games and Economic

---

[2] We adopt the following conventions for inequalities involving vectors: $\geqq$ means greater than or equal to in all components; $\geq$ means $\geqq$ and greater than in at least one component; $>$ means greater than in all components.

Behavior" as Lemmas (16:B) (= the "Theorem of the Supporting Plane" for convex bodies), (16:C), (16:D), (16:G). The relevant difference consists not in the slightly stronger form of Corollary 1 but in the algebraic nature of the proof of Stiemke's Theorem and the subsequent transition from Corollary 3 to the fundamental theorem for an arbitrary game.]

§4. CHARACTERIZATION OF POSSIBLE SOLUTION OF A SYMMETRIC GAME

When investigating the nature of the sets of optimal strategies for an m by n game $A = (a_{ij})$ it is convenient to consider the sets of mixed strategies as subsets of the cartesian spaces R_m and R_n respectively. Precisely, they are the fundamental simplexes Q_1 and Q_2, spanned by the unit vectors in each space. As is well known, the optimal strategies form <u>convex polyhedral</u> subsets O_1 and O_2 of Q_1 and Q_2, respectively, but not all such pairs of polyhedra can be obtained as the solutions of an m by n game. A simple characterization of the possible sets of optimal strategies for an arbitrary m by n game appears elsewhere in this Study ([2], [3]).

To state this result, we call a pure strategy <u>essential</u> if it occurs with positive probability in some optimal mixed strategy; otherwise it is called <u>superfluous</u>. We denote by e_1 and e_2 (s_1 and s_2) the number of essential (superfluous) pure strategies for the two players. Moreover, let d_1 and d_2 be the dimensions of O_1 and O_2 respectively. We call a $(d_1 - 1)$-dimensional bounding face of O_1 <u>interior</u> if it does not lie in a $(d_1 - 1)$-dimensional face of Q_1 with a similar definition for O_2; the number of interior bounding faces for the two players will be denoted by f_1 and f_2 and are defined to be 0 if the optimal strategies are unique.

THEOREM. The pair of convex polyhedra O_1 and O_2 are the optimal strategies for an m by n game with matrix A if and only if:

(1) $e_1 - d_1 = e_2 - d_2$

(2) $f_1 \leq s_2, \quad f_2 \leq s_1$.

The following theorem gives the characterization for the more restrictive class of symmetric games; it is readily recognized as a generalization of Kaplansky's theorem [4] that for an n by n symmetric game to be completely <u>mixed</u> (that is, for all n pure strategies to appear in every optimal strategy) it is necessary that n be odd.

THEOREM. The pair of convex polyhedra O_1 and O_2 are the optimal strategies for an n by n symmetric game with matrix S if and only if:

(1) the sets O_1 and O_2 are isomorphic (that is, the identity map of R_n onto R_n takes O_1 onto O_2),

(2) $e_1 - d_1 = e_2 - d_2$ is odd,

(3) $f_1 \leq s_2$, $f_2 \leq s_1$.

PROOF. The necessity of (1) is obvious. As for (2), let S_e be the <u>essential submatrix</u> composed of the rows and columns corresponding to pure strategies; since the value of the game is zero, we have (see page 13 of [3]):

$$e_1 - d_1 = e_2 - d_2 = \text{rank } S_e + 1 .$$

Hence the evenness of the rank of a skew-symmetric matrix implies the necessity of (2). Clearly (3) persists from the general theorem.

To prove the sufficiency of these conditions we prove the following:

LEMMA. Given a linear subspace L of R_m of even dimension $2k$ there exists a linear transformation of R_m <u>onto</u> L which can be described by a skew-symmetric matrix A in terms of a given orthogonal basis for R_m.

PROOF. Since the transform of a skew-symmetric matrix A by any orthogonal matrix T is skew-symmetric,

$$[-(T^{-1}AT)' = -(T^{-1}A'T) = T^{-1}(-A')T = T^{-1}AT]$$

we may assume that L is spanned by the first $2k$ unit vectors of the basis for R_m. Then if we let $B = \begin{pmatrix} 0 & -1 \\ 1 & 0 \end{pmatrix}$ the matrix

$$A = \left(\begin{array}{cccc|c} B & & & & \\ & B & & & \\ & & \ddots & & 0 \\ & & & B & \\ \hline & & 0 & & 0 \end{array}\right) ,$$

which is composed of k copies of B down the main diagonal and zero's elsewhere, describes the desired transformation. Geometrically, the transformation consists of a projection onto L followed by a rotation of the 2-dimensional planes which span L through an angle of $\pi/2$.

Following [3] we construct our game matrix S in the form:

$$S = \begin{pmatrix} S_e & S_2 \\ S_1 & 0 \end{pmatrix} .$$

If we let $[O_2]$ be the smallest linear space containing O_2 and let $[O_2]^{\perp}$ be its orthogonal complement in the space R_{e_2} spanned by the essential unit vectors, the dimension of $[O_2]^{\perp}$ is even by condition (2). Hence the Lemma assures the existence of a skew-symmetric matrix S_e with $[O_2]^{\perp}$ as its image space and $[O_2]$ as its null space.

The existence of S_1 and S_2 is assured by (3) as in the general theorem and by the duality of the inequalities involved $S_1 = - S_2'$. Thus all of the conditions of Lemma 6 of [3] are met and we have constructed a symmetric game with the given sets of optimal strategies.

BIBLIOGRAPHY

[1] STIEMKE, E., Math. Annalen, 76 (1915), pp. 340-342.

[2] BOHNENBLUST, H. F., KARLIN, S., SHAPLEY, L., "The Solutions of Discrete Two-person Games," this Study.

[3] GALE, D., SHERMAN, S., "Solutions of Finite Two-person Games," this Study.

[4] KAPLANSKY, I., "A Contribution to von Neumann's Theory of Games," Annals of Mathematics, Vol. 46 (1945), pp. 474-479.

D. Gale

H. W. Kuhn

A. W. Tucker

Princeton University

REDUCTIONS OF GAME MATRICES[1*]

D. Gale, H. W. Kuhn, and A. W. Tucker[2]

In attempting to solve games with a large number of pure strategies it is natural to group together strategies that are similar or are subject to some intrinsic connection such as symmetry. In many cases, one feels that the optimal probabilities within such a grouping of pure strategies can be fixed without regard to the game as a whole. Computationally, this is certainly desirable, since it replaces a set of strategies by a single new strategy and thus reduces the size of the game. In this note we shall investigate the possibility of such reductions; related results have been obtained independently by Seymour Sherman and Max Woodbury[3].

To make this notion more precise, let $A = (a_{ij})$ be an m by n payoff matrix and assume that, by partitioning the rows of A, we have decomposed it into matrices A_1 and A_2. If A_1 has m_1 rows, let P be a vector (one-row matrix) consisting of m_1 non-negative components whose sum is one. Then

$$A = \begin{bmatrix} A_1 \\ A_2 \end{bmatrix} \longrightarrow A^* = \begin{bmatrix} PA_1 \\ A_2 \end{bmatrix}$$

is called an <u>elementary first-player reduction</u>. A natural mapping of the mixed strategies X^* for A^* into the mixed strategies X for A is defined by:

$$X^* = (\lambda, x_{m_1+1}, \ldots, x_m) \longrightarrow X = (\lambda P, x_{m_1+1}, \ldots, x_m) .$$

THEOREM 1. *Let A^* result from A by an elementary first-player reduction. Then $v(A^*) \leq v(A)$. Optimal strategies X^* map into optimal strategies X if, and only if, $v(A^*) = v(A)$.*

---

[1]Presented to the American Mathematical Society – see abstract, Bulletin A.M.S. 55 (1949) p. 1045.

[2]Princeton and Stanford Universities, under contracts with the Office of Naval Research.

[3]Items [1] and [2] in the Bibliography appended.

*Accepted as a direct contribution to ANNALS OF MATHEMATICS STUDY No. 24.

PROOF. If X^* maps onto X we have the vector equality

$$X^*A^* = XA .$$

If we choose an optimal X^*, then the inequality follows immediately, while $v(A^*) = v(A)$ implies that X is optimal for A. Conversely, if X^* and X are optimal simultaneously, then the value of both games is the common minimum of the n components of $X^*A^* = XA$.

The content of this theorem is to assert that a solution of the reduced game for the first player yields a solution of the original game if, and only if, the elementary reduction does not change the value of the game. It also shows that such an elementary reduction is possible for any decomposition of the matrix A, merely by choosing a P depending on some optimal first player strategy for A. However, this generality entails impracticability and so we must modify our reductions if we wish to find applicable conditions. To this end, let the game payoff matrix $A = (a_{ij})$ be exhibited as an array of submatrices

$$A = \begin{bmatrix} A_{11} & A_{12} & \cdots & A_{1q} \\ A_{21} & A_{22} & \cdots & A_{2q} \\ \vdots & \vdots & & \vdots \\ A_{p1} & A_{p2} & \cdots & A_{pq} \end{bmatrix} ,$$

where each A_{kl} is a submatrix of m_k rows and n_l columns $(k = 1, \ldots, p;\ l = 1, \ldots, q)$. For each k let P_k be a vector (one-row matrix) consisting of m_k non-negative components whose sum is one. Form the matrix $B = (b_{kl})$, where each b_{kl} is the <u>minimum</u> of the n_l components of P_kA_{kl}. Then

$$A = (a_{ij}) \longrightarrow B = (b_{kl})$$

will be called a <u>first-player reduction</u>. It is clear that elementary reductions constitute a special case of reductions. Again there is a natural mapping of the first-player mixed strategies U for B into the first-player mixed strategies X for A, defined by:

$$U = (u_1, \ldots, u_p) \longrightarrow X = (u_1P_1, \ldots, u_pP_p) .$$

THEOREM 2. Let B result from A by a first-player reduction. Then $v(B) \leq v(A)$. Optimal strategies U map into optimal strategies X if $v(B) = v(A)$.

PROOF. If U maps onto X we have the vector inequality (inequality componentwise): $UB \leq XA$. If we choose an optimal U, then the inequality follows immediately, while $v(B) = v(A)$ implies that the corresponding X is optimal for A.

The following example shows that the condition of Theorem 2 is not necessary: Let

$$A = \begin{pmatrix} 1 & -1 \\ -1 & 1 \end{pmatrix},$$

$A_{11} = (1, -1)$, $A_{21} = (1, -1)$ and hence $P_1 = (1)$, $P_2 = (1)$. Then

$$B = \begin{pmatrix} -1 \\ -1 \end{pmatrix}$$

and $v(B) = -1 < 0 = v(A)$ but $U = (1/2, 1/2)$ is optimal for B and maps into $X = (1/2, 1/2)$ which is optimal for A.

Naturally we have the dual <u>second-player reduction</u> which is achieved by vectors Q_l (one-column matrices) consisting of n_l non-negative components whose sum is one. From these we form the matrix $C = (c_{kl})$, where each c_{kl} is the <u>maximum</u> of the m_k components of $A_{kl}Q_l$. Here the natural mapping is from second-player mixed strategies V for C into the second-player mixed strategies Y for A; Theorem 3 is exactly analogous to Theorem 2.

THEOREM 3. Let C result from A by a second-player reduction. Then $v(C) \geq v(A)$. Optimal strategies V map into optimal strategies Y if $v(C) = v(A)$.

In order to convert Theorems 2 and 3 into an applicable criterion, we prove the following general lemma.

LEMMA. Let $B = (b_{kl})$ and $C = (c_{kl})$ be two p by q payoff matrices such that

$$b_{kl} \leq c_{kl} \quad \text{for } k = 1, \ldots, p \text{ and } l = 1, \ldots, q .$$

Then $v(B) = v(C)$ if, and only if, the two games have common optimal strategies $U = (u_k)$ and $V = (v_l)$ such that

$$b_{kl} = c_{kl} \quad \text{whenever} \quad u_k v_l > 0 .$$

PROOF. Let $U = (u_k)$ and $V = (v_l)$ be common optimal strategies for the two games such that

$$b_{kl} = c_{kl} \quad \text{whenever} \quad u_k v_l > 0 .$$

Then

$$v(B) = \sum\sum u_k b_{kl} v_l = \sum\sum u_k c_{kl} v_l = v(C) .$$

On the other hand, let $U = (u_k)$ be any optimal first-player strategy for B, and let $V = (v_l)$ be any optimal second-player strategy for C. Then, since $c_{kl} = b_{kl}$ for all k, l,

$$\sum u_k c_{kl} \geq \sum u_k b_{kl} \geq v(B) \quad \text{for } k = 1, \ldots, p ,$$

$$v(C) \geq \sum c_{kl} v_l \geq \sum b_{kl} v_l \quad \text{for } l = 1, \ldots, q .$$

So, if $v(B) = v(C)$, U and V are also optimal strategies for C and B, respectively. Moreover,

$$\sum\sum u_k b_{kl} v_l = \sum\sum u_k c_{kl} v_l .$$

Therefore, since $b_{kl} \leq c_{kl}$ for all k, l,

$$b_{kl} = c_{kl} \quad \text{whenever} \quad u_k v_l > 0 .$$

This completes the proof of the lemma.

THEOREM 4. Let B and C be first- and second-player reductions of A. If the two games have common optimal strategies $U = (u_k)$ and $V = (v_l)$ such that

$$b_{kl} = c_{kl} \quad \text{whenever} \quad u_k v_l > 0 ,$$

then all optimal first-player (second-player) strategies for B(C) map into optimal first-player (second-player) strategies for A.

PROOF. Theorem 4 is a direct consequence of Theorems 2 and 3 and the lemma above. It should be noted that $b_{kl} = c_{kl}$ implies that their common value is $v(A_{kl})$. This suggests the following restatement of Theorem 4, useful for the applications.

THEOREM 5. Let B and C be first- and second-player reductions of A. If the two games have common optimal strategies $U = (u_k)$ and $V = (v_l)$ such that P_k and Q_l are optimal strategies for A_{kl} whenever $u_k v_l > 0$, then all optimal first-player (second-player) strategies for B(C) map into optimal first-player (second-player) strategies for A.

APPLICATIONS

(a) Domination. It is a well known fact that if one row of a payoff matrix dominates another row componentwise then the reduced game in which the dominated row has been deleted yields optimal strategies for the original game. This provides perhaps the simplest application of Theorem 5.

To make the situation precise, let $A = (a_{ij})$ be an m by n payoff matrix and assume that

$$a_{1j} \geq a_{2j} \quad \text{for} \quad j = 1, \ldots, n .$$

Then we set

$$A_{1j} = \begin{bmatrix} a_{1j} \\ a_{2j} \end{bmatrix}, \quad A_{kj} = (a_{k+1,j}) \quad \text{for} \quad j = 1, \ldots, n; \; k = 2, \ldots, m-1 .$$

$$P_1 = (1, 0), \; P_k = (1) \quad \text{for} \quad k = 2, \ldots, m-1 .$$

$$Q_j = (1) \quad \text{for} \quad j = 1, \ldots, n .$$

Then P_k and Q_j are optimal strategies for all of the subgames A_{kj} and both B and C are equal to A with the second row deleted. Hence Theorem 5 asserts that optimal first-player strategies for A can be obtained by solving this reduced game and filling in a 0 in the second component. It also asserts that the optimal second-player strategies coincide for A and the reduced game.

Naturally, the dual situation prevails when one column is dominated by another.

(b) Duplicated partial rows and columns. Again let $A = (a_{ij})$ be an m by n payoff matrix and assume that the first r rows of A have the same last $n - s$ elements and the first s columns have the same last $m - r$ elements. Then we decompose A thus

$$A = \begin{bmatrix} A_{11} & A_{12} & \cdots & A_{1,n-s+1} \\ A_{21} & a_{r+1,s+1} & \cdots & a_{r+1,n} \\ \cdot & \cdot & & \cdot \\ \cdot & \cdot & & \cdot \\ \cdot & \cdot & & \cdot \\ A_{m-r,1} & a_{m,s+1} & \cdots & a_{mn} \end{bmatrix} .$$

Here A_{k1} and A_{1l} are one-row and one-column matrices for $k = 2, \ldots, m - r + 1; \; l = 2, \ldots, n - s + 1$. The reduction is effected by setting P_1 and Q_1 equal to optimal strategies for A_{11} while $P_k = Q_l = (1)$ $(k = 2, \ldots, m - r + 1; \; l = 2, \ldots, n - s + 1)$. Then the P_k and Q_l are optimal for all of the subgames A_{kl} and hence the first- and second-player reductions are the same; Theorem 5 asserts that optimal strategies for this reduced game map into optimal strategies for A.

(c) The essential game decomposition. For this application, we use the same decomposition used in (b) although we do not assume that the one-row and one-column games are identical. We assume, in addition, optimal strategies P_1 and Q_1 for A_{11} in which all the components are positive such that the first pure strategy is an optimal first-player strategy for B and the first pure strategy is an optimal second-player strategy for C. Then the lead element in both matrices is a saddle point and the hypothesis of Theorem 5 is again satisfied. Results of Bohnenblust, Karlin, Shapley, Gale and Sherman show that this reduction to a saddle-point matrix is always possible, the maximal sets of r rows and s columns being unique and defining the essential game.

(d) Symmetrization. The symmetrization proposed by the authors in the preceding paper yields an interesting theoretical application. There an arbitrary m by n game $A = (a_{ij})$ is "symmetrized" to the game described by the matrix

$$S = \left[\begin{array}{c|c|c} 0 & A & -1 \\ \hline -A' & 0 & 1 \\ \hline 1 & -1 & 0 \end{array}\right],$$

where $-A'$ is the negative transpose of A, ± 1 are vectors composed of the requisite number of ± 1's and the 0's stand for zero matrices which fill out S. Then, if $x = (x_1, \ldots, x_n)$ and $y = (y_1, \ldots, y_n)$ are optimal mixed strategies for A, we obtain the reduction by setting $P_1 = Q_1 = x$, $P_2 = Q_2 = y$, and $P_3 = Q_3 = (1)$, using the partition indicated by the dotted lines. Since P_k and Q_l are optimal for all of the subgames A_{kl}, the first- and second-player reductions are the same, i.e.,

$$B = C = \begin{bmatrix} 0 & v & -1 \\ -v & 0 & 1 \\ 1 & -1 & 0 \end{bmatrix},$$

where v is the value of A. If $v \leq 0$, then the second strategies of the reduced game provide a saddle point. If $v > 0$, the optimal strategy is unique and equals $(\frac{1}{2+v}, \frac{1}{2+v}, \frac{v}{2+v})$. If $v = 1$, our reduced game is the common game "Stone, Paper, Scissors."

(e) Submatrices within each of which column totals are equal and row totals are equal. Suppose a payoff matrix A can be partitioned into submatrices A_{kl} each of which has all its column totals equal and all its row totals equal. (By a column or row total in any matrix we mean the sum of all the entries in a particular column or row of that matrix.) Then, taking vectors P_k in which each component is $1/m_k$ and vectors Q_l in which each component is $1/n_l$, we obtain first- and second-player reductions B and C that are identical because $b_k = c_k$ = average of all the entried in A_{kl}. In this case, by Theorem 4, all optimal strategies for B = C map into optimal strategies for A.

An illustration of this reduction is furnished by the following somewhat simplified version of the deployment problem of "Colonel Blotto." Colonel Blotto is required, as a test of his military competence, to make a strategic plan for the deployment of three military units among three mountain passes against two enemy units. Blotto scores one point for each pass in which he has more units than the enemy and one point for each enemy unit overpowered, and loses one point in each reverse instance. The payoff matrix A, below at left, is 10 by 6 – corresponding to the possible opposing deployments (e.g., 210 means two units at the first pass, one at the second, and none at the third).

A	200	020	002	110	101	011
300	3	0	0	1	1	-1
030	0	3	0	1	-1	1
003	0	0	3	-1	1	1
210	1	-1	1	2	2	0
201	1	1	-1	2	2	0
120	-1	1	1	2	0	2
021	1	1	-1	2	0	2
102	-1	1	1	0	2	2
012	1	-1	1	0	2	2
111	0	0	0	1	1	1

B = C	2	1 & 1
3	1	1/3
2 & 1	1/3	4/3
1,1,1	0	1

This payoff matrix A can be partitioned into 3 by 2 submatrices (as indicated by the broken lines) each of which has all its column totals equal and all of its row totals equal. The corresponding reduced matrix B = C, above at right, consists of entries which are the averages of the submatrices of A. Optimal strategies (3/5, 2/5, 0) and (3/5, 2/5) for B = C map into

$$(1/5,\ 1/5,\ 1/5;\ 1/15,\ 1/15,\ 1/15,\ 1/15,\ 1/15,\ 1/15;\ 0)$$

and

$$(1/5,\ 1/5,\ 1/5;\ 2/15,\ 2/15,\ 2/15)$$

for A. The common game-value of A and B = C is 11/15.

BIBLIOGRAPHY

[1] SHERMAN, S., "Games and Sub-games," Abstract No. 327t, Bulletin of the Amer. Math. Soc. 56 (1950), p. 347 (to be published in the Proceedings of the Amer. Math. Soc. in 1951).

[2] WOODBURY, M., "On Games Whose Matrices are the Sums of the Matrices of Other Games," Abstract No. 122t, Bulletin of the Amer. Math. Soc. 56 (1950), p. 160.

D. Gale

H. W. Kuhn

A. W. Tucker

Princeton University

A SIMPLIFIED TWO-PERSON POKER*

H. W. Kuhn[1]

A fascinating problem for the game theoretician is posed by the common card game, Poker. While generally regarded as partaking of psychological aspects (such as bluffing) which supposedly render it inaccessible to mathematical treatment, it is evident that Poker falls within the general theory of games as elaborated by von Neumann and Morgenstern [1]. Relevant probability problems have been considered by Borel and Ville [2] and several variants are examined by von Neumann [1] and by Bellman and Blackwell [3].

As actually played, Poker is far too complex a game to permit a complete analysis at present; however, this complexity is computational and the restrictions that we will impose serve only to bring the numbers involved within a reasonable range. The only restriction that is not of this nature consists in setting the number of players at two. (The games considered in [1] and [3] also require this condition.) The simplifications, though radical, enable us to compute <u>all</u> optimal strategies for both players. In spite of these modifications, however, it seems that Simplified Poker retains many of the essential characteristics of the usual game.

An <u>ante</u> of one unit is required of each of the two players. They obtain a fixed <u>hand</u> at the beginning of a play by drawing one card apiece from a pack of three cards (rather than the $\binom{52}{5} = 2{,}598{,}960$ hands possible in Poker) numbered 1, 2, 3. Then the players choose alternatively either to <u>bet</u> one unit or <u>pass</u> without betting. Two successive bets or passes terminate a play, at which time the player holding the higher card wins the amount wagered previously by the other player. A player passing after a bet also ends a play and loses his ante.

Thus thirty possible plays are permitted by the rules. First of all, there are six possible deals; for each deal the action of the players may follow one of five courses which are described in the following diagram:

[1] Princeton University under a contract with the Office of Naval Research.
* Accepted as a direct contribution to ANNALS OF MATHEMATICS STUDY NO. 24.

	First Round Player I	First Round Player II	Second Round Player I	Payoff
(1)	pass	pass		1 to holder of higher card
(2)		bet	pass	1 to player II
(3)			bet	2 to holder of higher card
(4)	bet	pass		1 to player I
(5)		bet		2 to holder of higher card

We code the pure strategies available to the players by ordered triples (x_1, x_2, x_3) and (y_1, y_2, y_3) for players I and II, respectively ($x_i = 0, 1, 2$; $y_j = 0, 1, 2, 3$). The instructions contained in x_i are for card i and are deciphered by expanding x_i in the binary system, the first figure giving directions for the first round of betting, the second giving directions for the second, with 0 meaning pass and 1 meaning bet. For example, $(x_1, x_2, x_3) = (2, 0, 1) = (10, 00, 01)$ means player I should bet on a 1 in the first round, always pass with a 2 and wait until the second round to bet on a 3.

Similarly, to decode y_j, one expands in the binary system, the first figure giving directions when confronted by a pass, the second when confronted by a bet, with 0 meaning pass and 1 meaning bet. Thus $(y_1, y_2, y_3) = (2, 0, 1) = (10, 00, 01)$ means that player II should pass except when holding a 1 and confronted by a pass or holding a 3 and confronted by a bet.

In terms of this description of the pure strategies, the payoff to player I is given by the following scheme:

x_i \ y_j	0 = 00	1 = 01	2 = 10	3 = 11
0 = 00	± 1	± 1	-1	-1
1 = 01	± 1	± 1	± 2	± 2
2 = 10	1	± 2	1	± 2

where the ambiguous sign is + if $i > j$ and - if $i < j$.

From the coding of the pure strategies it is clear that player I has 27 pure strategies while player II has 64 pure strategies. Fortunately Poker sense indicates a method of reducing this unwieldy number of strategies.

Obviously, no player will decide either to bet on a 1 or to pass with a 3 when confronted by a bet. For player I (II) this heuristic argument recognizes the domination of certain rows (columns) of the game matrix by other rows (columns). It is well known that we may drop the dominated rows (dominating columns) without changing the value of the game and that any optimal strategy for the matrix thus reduced will be optimal for the original game. However, since the domination is not proper, these pure strategies could appear in some optimal mixed strategy. For the careful solver who may wish to find all of the optimal strategies, complementary arguments may be made to show that the pure strategies dropped are actually superfluous in this game. After we have found at least one of the optimal strategies for each player we shall give an indication of these arguments.

Now that these strategies have been eliminated new dominations appear. First, we notice that if player I holds a 2 he may as well pass in the first round, deciding to bet in the second if confronted by a bet, as bet originally. On either strategy he will lose the same amount if player II holds a 3; on the other hand, player II may bet on a 1 if confronted by a pass but certainly will not if confronted by a bet. Secondly player II may as well pass as bet, when holding a 2 and confronted by a pass since player I will now answer a bet only when he holds a 3.

We are now in a position to describe the game matrix composed of those strategies not eliminated by the previous heuristic arguments. Each entry is computed by adding the payoffs for the plays determined by the six possible deals for each pair of pure strategies and thus this matrix is six times the actual game matrix.

(x_1,x_2,x_3) \ (y_1,y_2,y_3)	(0,0,3)	(0,1,3)	(2,0,3)	(2,1,3)
(0,0,1)	0	0	-1	-1
(0,0,2)	0	1	-2	-1
(0,1,1)	-1	-1	1	1
(0,1,2)	-1	0	0	1
(2,0,1)	1	-2	0	-3
(2,0,2)	1	-1	-1	-3
(2,1,1)	0	-3	2	-1
(2,1,2)	0	-2	1	-1

One easily verifies that the following mixed strategies are optimal for this game matrix (and hence for Simplified Poker):

Player I:

(A)	2/3	(0,0,1)	+	1/3	(0,1,1)			
(B)	1/3	(0,0,1)	+	1/2	(0,1,2)	+	1/6	(2,0,1)
(C)	5/9	(0,0,1)	+	1/3	(0,1,2)	+	1/9	(2,1,1)
(D)	1/2	(0,0,1)	+	1/3	(0,1,2)	+	1/6	(2,1,2)
(E)	2/5	(0,0,2)	+	7/15	(0,1,1)	+	2/15	(2,0,1)
(F)	1/3	(0,0,2)	+	1/2	(0,1,1)	+	1/6	(2,0,2)
(G)	1/2	(0,0,2)	+	1/3	(0,1,1)	+	1/6	(2,1,1)
(H)	4/9	(0,0,2)	+	1/3	(0,1,1)	+	2/9	(2,1,2)
(I)	1/6	(0,0,2)	+	7/12	(0,1,2)	+	1/4	(2,0,1)
(J)	5/12	(0,0,2)	+	1/3	(0,1,2)	+	1/4	(2,1,1)
(K)	1/3	(0,0,2)	+	1/3	(0,1,2)	+	1/3	(2,1,2)
(L)	2/3	(0,1,2)	+	1/3	(2,0,2)			

Player II:

1/3	(0,0,3)	+	1/3	(0,1,3)	+	1/3	(2,0,3)
2/3	(0,0,3)	+	1/3	(2,1,3)			

These strategies yield the value of Simplified Poker as $-1/18$.

As an example of the complementary arguments which assure us that no solutions are lost by discarding dominated rows and dominating columns, consider the pure strategies of the form $(1, x_2, x_3)$ which we have eliminated for player I. The verbal arguments assure us that player I will do at least as well by using $(0, x_2, x_3)$ no matter what mixed strategy II uses and irrespective of the deal. However, if I is dealt card 1 and II is dealt card 3, and if II plays either of his optimal strategies then player I loses two units when he plays $(1, x_2, x_3)$ while he loses but one unit when he plays $(0, x_2, x_3)$. Thus, since all of the pure strategies $(0, x_2, x_3)$ are <u>essential</u>, player I's expectation is less than the value of the game when he plays $(1, x_2, x_3)$ against II's optimal strategies and we see that the pure strategies $(1, x_2, x_3)$ are <u>superfluous</u>. (These terms are used as defined by Gale and Sherman in [4].)

The rank of the essential matrix is easily computed to be 3 (the sum of the first and last columns is equal to the sum of the center two columns), hence, by results of Bohnenblust, Karlin, and Shapley [5], Gale and Sherman [4], players I and II have precisely 6 and 2 linearly independent optimal strategies, respectively. A simple application of the work of Shapley and Snow [6] proves that all of the optimal strategies given above are <u>basic</u>. Therefore we know immediately that we have found all of the basic optimal strategies for player II. The corresponding result for player I is proved by considering the remaining <u>kernels</u> in the sense of [6]. This calculation is facilitated by the fact that, since the essential game was obtained by dominations, all solutions of the essential game can be extended to solutions of the full game and hence we need only consider kernels within the essential game. Moreover, the knowledge of the full set of solutions for II enables I to eliminate all 2 by 2 kernels except those involving the first and last column; there can be no 4 by 4 kernels since the essential game matrix has rank 3 and the value of the game is different from zero. Thus we verify that all optimal strategies are convex linear combinations of the strategies given above.

A striking simplification of the solution is achieved if we return to the extensive form of the game. We do this by introducing behavior parameters to describe the choices remaining available to the players after we have eliminated the superfluous strategies. We define:

Player I:

α = probability of bet with 1 in first round.
β = probability of bet with 2 in second round.
γ = probability of bet with 3 in first round.

Player II:

ξ = probability of bet with 1 against a pass.
η = probability of bet with 2 against a bet.

In terms of these parameters, player I's basic optimal strategies fall into seven sets:

Basic Strategies	(α, β, γ)
A	(0, 1/3, 0)
C	(1/9, 4/9, 1/3)
E	(2/15, 7/15, 2/5)
B,D,F,G	(1/6, 1/2, 1/2)
H	(2/9, 5/9, 2/3)
I,J	(1/4, 5/12, 3/4)
K,L	(1/3, 2/3, 1)

Thus, in the space of these behavior parameters, the five dimensions of optimal mixed strategies for player I collapse onto the one parameter family of solutions:

$$\alpha = \gamma/3$$
$$\beta = \gamma/3 + 1/3$$
$$0 \leq \gamma \leq 1$$

These may be described verbally by saying that player I may pass on a 3 in the first round with arbitrary probability, but then he must bet on a 1 in the first round one third as often, while the probability with which he bets on a 2 in the second round is one third more than the probability with which he bets on a 1 in the first round.

On the other hand, we find that player II has the single solution:

$$(\xi, \eta) = (1/3, 1/3) \; ,$$

which instructs him to bet one third of the time when holding a 1 and confronted by a pass and to bet one third of the time when holding a 2 and confronted by a bet.

The presence of <u>bluffing</u> and <u>underbidding</u> in these solutions is noteworthy (<u>bluffing</u> means betting with a 1; <u>underbidding</u> means passing on a 3). All but the extreme strategies for player I, in terms of the behavior parameters, involve both bluffing and underbidding while player II's single optimal strategy instructs him to bluff with constant probability 1/3 (underbidding is not available to him). These results compare favorably with presence of bluffing in the von Neumann example, while bluffing is not available to player II in the continuous variant considered by Bellman and Blackwell.

The sensitive nature of bluffing and underbidding in this example is exposed by varying the ratio of the bet to the ante. Consider the games described by the same rules in which the ante is a positive real number a and the bet is a positive real number b. We will state the solutions in terms of the behavior parameters without proof. This one parameter family of games falls naturally into four intervals:

<u>Case 1: $0 < b < a$</u>

Player I: $(\alpha, \beta, \gamma) = \left(\frac{b}{2a + b}, \frac{2a}{2a + b}, 1\right)$

Player II: $(\xi, \eta) = \left(\frac{b}{2a + b}, \frac{2a - b}{2a + b}\right)$

Remarks: Both players have a unique optimal mode of behavior. Player I never underbids and bluffs with probability $\frac{b}{2a + b}$; player II bluffs with probability $\frac{b}{2a + b}$. The value of this game is $-\frac{b^2}{6(2a + b)}$.

<u>Case 2: $0 < b = a$</u>

This is our original game. The value is $-\frac{b}{18}$.

<u>Case 3: $0 < a < b < 2a$</u>

Player I: $(\alpha, \beta, \gamma) = \left(0, \frac{2a - b}{2a + b}, 0\right)$

Player II: $\xi = \frac{b}{2a + b}$

$$\frac{b}{2a + b} \leq \eta \leq \frac{a + b}{2a + b}$$

Remarks:: Player I never bluffs and always underbids; player II bluffs with probability $\frac{b}{2a + b}$. The value of this game is $-\frac{b}{6}\left(\frac{2a - b}{2a + b}\right)$.

Case 4: $0 < 2a = b$

This game has a saddle point in which player I never bluffs, always underbids and never bets on a 2 while player II bets only and always with a 3. The strategy for I is unique while II can vary his strategy considerably. The value of this game is 0.

It is remarkable that player I has a negative expectation for a play, i.e., a disadvantage that is plausibly imputable to his being forced to take the initiative. (Compare von Neumann's variant (c), in which the possession of the initiative seems to be an advantage.) It is also noteworthy that Simplified Poker, which was not constructed with an eye to solutions, but rather as a modification of an actual game, has many solutions notwithstanding the fact that games with unique optimal strategies are dense in the space of all games.

In conclusion, it is hoped that these considerations will prove instructive in a qualitative way and contribute in some small measure to the casuistry of game solving.

BIBLIOGRAPHY

[1] MORGENSTERN, O., von NEUMANN, J.,"The Theory of Games and Economic Behavior," 2nd ed., (1947) Princeton University Press, Princeton.

[2] BOREL, E., and collaborators,"Traité du Calcul des Probabilités et de ses Applications," Vol. IV, 2, Chapt. 5, (1938) Paris.

[3] BELLMAN, R., BLACKWELL, D.,"Some Two-Person Games Involving Bluffing," Proc. Nat. Acad. Sci., Vol. 35, p. 600 (1949).

[4] GALE, D., SHERMAN, S.,"Solutions of Finite Two-Person Games," this study.

[5] BOHNENBLUST, H. F., KARLIN, S., SHAPLEY, L.,"The Solutions of Discrete Two-Person Games," this study.

[6] SNOW, R. M., SHAPLEY, L.,"Basic Solutions of Discrete Games," this study.

H. W. Kuhn

Princeton University

A SIMPLE THREE-PERSON POKER GAME[1 *]

J. F. Nash[2] and L. S. Shapley

§1. INTRODUCTION

In the study of games Poker, in its varied forms, has become a popular source of models for mathematical analysis. Various simple Pokers have been investigated by von Neumann,[3] Bellman and Blackwell,[4] and Kuhn.[5,6] Our paper is the first to consider a three-person model. This version has just two kinds of hands, no drawing or raising, and only one size of bet. We suppose that the game is non-cooperative and solve for "equilibrium points." The game turns out to have a well-defined value if the ante does not exceed the amount of the bet, or is more than four times the bet; but no value for at least two transition cases in between.

To cut down the magnitude of the computational task we use "behavior coefficients" in place of mixed strategies. This is an effective technique for a large class of games in extensive form.

§2. THE SOLUTION OF AN n-PERSON GAME

A definition for the solution of an n-person game, $n > 2$, based on the principle of coalition, has been developed by von Neumann and Morgenstern.[7] It is unfortunately weak in its ability to predict the actions of the players, or to ascribe a value to the game. It is most naturally applicable to games (or economic situations) in which the players are free to offer or accept side payments (outside the mechanism of the game itself) in

---

[1]This work was supported in part by the Office of Naval Research.

[2]AEC Fellow.

[3]J. von Neumann and O. Morgenstern, "Theory of Games and Economic Behavior," 2nd ed., Princeton, 1947; pp. 186-219.

[4]Proc. N. A. S. 35 (1949), pp. 600-605.

[5]In this volume.

[6]E. Borel considers some simple two-person betting models in the course of an analysis of the probabilities of actual Poker games, in his "Traité du Calcul des Probabilités, Paris, 1938; IV, 2, pp. 91-97.

[7]Op. cit., Chapter VI.

*Accepted as a direct contribution to ANNALS OF MATHEMATICS STUDY No. 24.

return for cooperation during the play. The customary ethics of Poker suggest that a non-cooperative solution concept, not recognizing side payments and pre-play agreements, would be better for our present purpose. We therefore define:

An equilibrium point (or EP) is a set of strategy choices, pure or mixed, of the n players, with the property that no player can improve his expectation by changing his own choice, the others being held fixed.

If it happens that each player's expectation is the same in all equilibrium points, then we call the n-tuple of these expectations the value of the game.

In a two-person, zero-sum game the equilibrium points are just the minimax points, and describe the solution in the usual sense. It has been shown that finite games always possess EP.[8] They do not necessarily have values; nor are the strategies used in different EP of the same game necessarily interchangeable.

§3. THE RULES OF THE GAME

The deck contains just two kinds of cards, "High" and "Low," in equal numbers. One card is dealt at random to each of the three players. The deck is so large that the eight possible deals occur with equal probability. Each player antes an amount a. The first player has the option of opening the bidding with a bet b, or of passing. If he passes, the second player has the same opportunity; then the third. When any player has opened, the other two, in rotation, have the choice of calling with a bet b, or of folding (dropping out), thereby forfeiting the ante money. The payoff rule: If no one opened (three consecutive passes), the players retrieve their antes. Otherwise, the players betting compare cards, and the one with the highest wins the entire accumulation of bets and antes (the pot). In case of a tie, the winners divide the pot equally.

There are 13 possible sequences of bids. We may represent them:

BBB BPB PBBB PBPB PPBBB PPBPB PPP
BBP BPP PBBP PBPP PPBBP PPBPP

letting "B" stand for "open" and "call," "P" for "pass" or "fold." With the eight possible deals, there are 104 different plays of the game that can occur. The possible payoffs are seen to be:

$$(2a, -a, -a) \quad (a/2, a/2, -a)$$
$$(2a + b, -a, -a - b) \quad (a/2 + b/2, a/2 + b/2, -a - b)$$
$$(2a + 2b, -a - b, -a - b) \quad (0, 0, 0)$$

---

[8] J. Nash, Proc. N. A. S. 36 (1950), pp. 48-49.

and their permutations. Clearly it is only the ratio of a to b that is significant, but we shall retain the separate symbols in the hope that mental verifications by the reader of such statements as the above will be made easier.

§ 4. BEHAVIOR COEFFICIENTS

By a calculation which we do not detail, we find that the three players have respectively 81, 100, and 256 pure strategies. If we forbid those that involve folding on a high card (see below, §7), these numbers reduce to 18, 20, and 32. The equilibrium points are then to be sought in a space of 17 + 19 + 31 = 67 dimensions, the product of the three mixed-strategy simplexes.

However, as commonly occurs in games with several moves, there is great redundancy in this representation. Distinct mixed strategies exist which prescribe identical behaviors for the player in question. This equivalence induces a natural projection of his simplex into a convex polytope of much lower dimension, with each pure strategy going into a distinct extreme point of the polytope. We shall discover that the players have only eight essential dimensions apiece; or five apiece if we outlaw folding on a high card.

The natural way to achieve this economy is by avoiding the description of behavior as a probability mixture of the pure strategies, and instead considering on-the-spot randomizations during the course of play.[9] There are eight situations which require a decision that can face each player; and the decision is always to make or not to make a bet. We therefore introduce as the <u>behavior coefficients</u> the probabilities of betting in the different situations:

Player 1

He holds	High	Low
Faced with:	he bets with probability:	
--	α	β
PBB	o	π
PBP	ρ	σ
PPB	τ	υ

Player 2

He holds	High	Low
Faced with:	he bets with probability:	
B	γ	δ
P	ε	ζ
PPBB	φ	χ
PPBP	ψ	ω

Player 3

He holds	High	Low
Faced with:	he bets with probability:	
BB	η	θ
BP	ι	κ
PB	λ	μ
PP	ν	ξ

BEHAVIOR COEFFICIENTS

[9]This procedure would not be legitimate if each player's information were not monotone increasing.

Every mixed strategy can be completely represented, as far as its effect in the game is concerned, by an octet of values of the appropriate player's behavior coefficients.

§5. IRRELEVANCE

It can happen that if certain of the coefficients take on extreme values (i.e., 0 or 1), then the situations to which other coefficients apply can never arise. For example, if $\alpha = 1$ and $\beta = 1$ then the values of ε and ζ tell nothing about the behavior of Player 2. To keep the representation of behavior unique, we assign the conventional value 1 to an irrelevant coefficient if it refers to a high card, 0 if it refers to a low card This amounts to identifying certain vertices of the 24-dimensional cube of behavior probabilities, and does not lower dimension.

§6. DISCRIMINANTS

The expected payoffs to the three players will be certain multilinear functions[10]

$$P_1(\alpha, \beta, \ldots, \omega),\ P_2(\alpha, \beta, \ldots, \omega),\ P_3(\alpha, \beta, \ldots, \omega)$$

of the behavior coefficients, with terms of degree as high as five. It would be tedious, as well as unnecessary, to attempt to give the explicit functions here. In their place we shall work with the <u>discriminants</u>, $\Delta_\alpha, \Delta_\beta, \ldots, \Delta_\omega$, defined by:

$$\Delta_u = 16 \frac{\partial}{\partial u} P_{k(u)}(\alpha, \beta, \ldots, \omega), \quad u = \alpha, \beta, \ldots, \omega ;$$

where it is the $k(u)$-th player who controls u. (The factor of 16 clears out the fractions arising from the random deal and the divided pots.) Directly from the definition of equilibrium, we have, at any EP:

$$\text{(C)} \qquad \begin{cases} u = 0 \implies \Delta_u \leq 0 \\ 0 < u < 1 \implies \Delta_u = 0 \\ u = 1 \implies \Delta_u \geq 0 \end{cases} \qquad u = \alpha, \beta, \ldots, \omega .$$

A set of coefficient-values satisfying (C) does not necessarily constitute an EP, since the possibility that a player would increase his expectation by varying two or more of his coefficients simultaneously is not excluded by (C). Our method of solution will be to show that among the possible EP's

---

[10] In the mixed strategies they would be trilinear functions.

only one satisfies (C). By the existence theorem[11] this one must be the unique EP.

§ 7. DOMINATIONS

The values in the solution of 9 of the 24 coefficients can be fixed immediately by observing that to drop out with a high card is always definitely injurious to a player's expectation. Thus we have ...

$$\ldots \boxed{\gamma = \eta = \iota = \lambda = o = \rho = \tau = \varphi = \psi = 1}\,.$$

The value of ν can not quite be determined in the same way, for Player 3 might conceivably find opening on High no more profitable than passing out the hand. However, it is easily seen that if an equilibrium point exists with $\nu < 1$, then $\nu = 1$ together with the same set of other values is also an EP. Therefore we may assume ... $\boxed{\nu = 1}$. It will turn out that the unusual circumstances (i.e., $\beta = \zeta = 1$; $\alpha, \varepsilon < 1$) which permit $\nu < 1$ do not occur in any EP.

Before proceeding, we restrict ourselves to the case $b \geq a$. Later, by a process of continuation, we shall find EP's for smaller bet sizes. But the demonstration of the existence of a value to the game requires that <u>all</u> EP's be known, and to prove completeness is too complicated for $b < a$.

We now show, in a more elaborate argument, that $b \geq a$ entails $\beta = 0$. If Player 1 opens on Low then he must expect to lose $a + b$ in .75 of the deals, and gains at most $2a$ in the remaining .25. This expectation of at most $-\frac{1}{4}(a + 3b)$ must be compared with that of at least $-a$ he can obtain by not betting at all (i.e., $\beta = \pi = \sigma = \upsilon = 0$). Since we have assumed $b \geq a$, a behavior involving opening on Low (i.e., $\beta > 0$) is possible in an EP only if conditions are most favorable to that policy. That is, b and a must be equal, and

(i) Player 1 must always win the amount 2a by opening in the low-low-low deal (i.e., $\delta = \kappa = 0$);

(ii) He must never be allowed to recover his ante when he passes (i.e., either $\xi = 1$ or $\varepsilon = \zeta = 1$).

These conditions have a decisive effect on α. We may estimate Δ_α by the following tabulation of Player 1's payoff:

[11] Nash, loc. cit.

Deal	$\alpha = 1$	$\alpha = 0$
HHH	0	0
HHL	at most $(a+b)/2$	at least $a/2$
HLH	$a/2$	at least $a/2$
HLL	$2a$	at least $2a + b$

We conclude that $\Delta_{\alpha} < -b$ and $\alpha = 0$. It is now easily verified that $\Delta_{\kappa} = 3\beta a$. But, with κ already zero by (i), there can be no equilibrium with $\Delta_{\kappa} > 0$ (condition (C)).[12] Thus, even after assuming most favorable conditions for β, we are led to the conclusion ... $\boxed{\beta = 0}$.

§8. FURTHER REDUCTIONS

With $\beta = 0$, it is easily calculated that:

$$\Delta_{\delta} = -4\alpha b$$
$$\Delta_{\theta} = -2\alpha(1 + \delta)b$$
$$\Delta_{\kappa} = -2\alpha(1 - \delta)b \ .$$

If $\alpha > 0$ then these three discriminants are strictly negative.[13] But if $\alpha = 0$ then δ, θ, and κ become irrelevant. In any case ... $\boxed{\delta = \theta = \kappa = 0}$.

Continuing, we have:

$$\Delta_{\pi} = 2(\zeta\mu a - \zeta b - \varepsilon\mu b - \varepsilon b)$$
$$\Delta_{\chi} = 2\bar{\zeta}(\xi \upsilon a - \upsilon b - \bar{\alpha}\xi b - \bar{\alpha} b) \ .$$

(We use the bar to denote complementary probabilities: $\bar{\alpha} = 1 - \alpha$, etc.) If either of these discriminants is to be non-negative then several of the coefficients are forced to assume extreme values, and a and b must be equal. When these restrictions are applied to the other discriminants a contradiction for each case is soon reached. The process is similar in form to the proof of $\beta = 0$, as given above, and we shall not burden this account with the details. The conclusion is ... $\boxed{\pi = \chi = 0}$.

A succession of results can be established in like manner, by deducing contradictions from the alternative hypotheses We list them in the order in which we found the arguments to go through most easily.

---

[12]The corresponding verbal argument: when confronted by BP, Player 3 knows that both opponents hold Low; hence a call is always profitable for him.

[13]For Δ_{κ} we argue:

$$\Delta_{\delta} < 0 \implies \delta = 0 \implies \Delta_{\kappa} < 0 \ .$$

$\alpha > 0;$	$\xi < \frac{2}{3};$	$\nu = \omega = 0;$	$\varepsilon < 1;$	$\sigma < 1,\ \mu = 0;$
$\zeta < 1;$	$\xi > 0;$	$\varepsilon > 0;$	$\alpha < 1;$	$\sigma = 0.$

Again we omit the details, which are rather lengthy and not particularly interesting.

§ 9. THE SOLUTION FOR $b \geq a$

There remain at this juncture two systems of equations, differing in the way in which they involve ζ, which might possibly determine an EP. Together with the inequalities which they entail, they are:

$$(\mathrm{I})\ \begin{cases} \Delta_\alpha = 0, & 0 < \alpha < 1 \\ \Delta_\varepsilon = 0, & 0 < \varepsilon < 1 \\ \zeta = 0, & \Delta_\zeta \leq 0 \\ \Delta_\xi = 0, & 0 < \xi < 2/3 \end{cases} \qquad (\mathrm{II})\ \begin{cases} \Delta_\alpha = 0, & 0 < \alpha < 1 \\ \Delta_\varepsilon = 0, & 0 < \varepsilon < 1 \\ \Delta_\zeta = 0, & 0 \leq \zeta < 1 \\ \Delta_\xi = 0, & 0 < \xi < 2/3 . \end{cases}$$

The four discriminants in question are:

$$\Delta_\alpha = (a+b)\bar{\xi}\,\bar{\varepsilon} + (4a+2b)\bar{\xi}\,\bar{\zeta} - b\bar{\varepsilon} + b\bar{\zeta} - 3b ,$$

$$\Delta_\varepsilon = (a+b)\bar{\xi}\bar{\alpha} + (4a+2b)\bar{\xi} - b\bar{\alpha} - 2b ,$$

$$\Delta_\zeta = -2a\bar{\xi}\bar{\alpha} - 2a\bar{\xi} - 4b\bar{\alpha} + 6a - 2b ,$$

$$\Delta_\xi = -2(a+b)(\bar{\alpha}\,\bar{\varepsilon} + \bar{\alpha}\,\bar{\zeta} + \bar{\varepsilon}) + 4a\bar{\zeta} .$$

The solution of system (I) is:

$$\alpha = \varepsilon = 2 - S,\ \zeta = 0,\ \xi = 1 - \frac{b}{(a+b)S};\ S = \sqrt{\frac{3a+b}{a+b}} .$$

The inequalities are satisfied in the range:

R_I $\qquad 0 < a/b \leq A_1 = 0.7058 \ldots .$[14]

If a/b exceeds A_1 then Δ_ζ becomes positive.

The equations of (II) give complicated expressions for α, ε, ζ, and ξ. The inequalities are satisfied in the range:

R_{II} $\qquad A_1 \leq a/b \leq 1 .$

If a/b is less than A_1, then ζ becomes negative. On the other hand, none of the inequalities is violated immediately if a/b is allowed to exceed 1. At the endpoints of R_{II} the numerical values are:

---

[14] A_1 is the positive root of $9A^4+18A^3+3A^2-10A-3$.

at $a/b = A_1$: $\alpha = .6482$ $\varepsilon = .6482$ $\zeta = 0$ $\xi = .5664$;

at $a/b = 1$: $.3084$ $.8257$ $.0441$ $.6354$.

As may be seen in Figure 1A, the coefficients are practically linear in R_{II}, and the connection at A_1 is continuous.

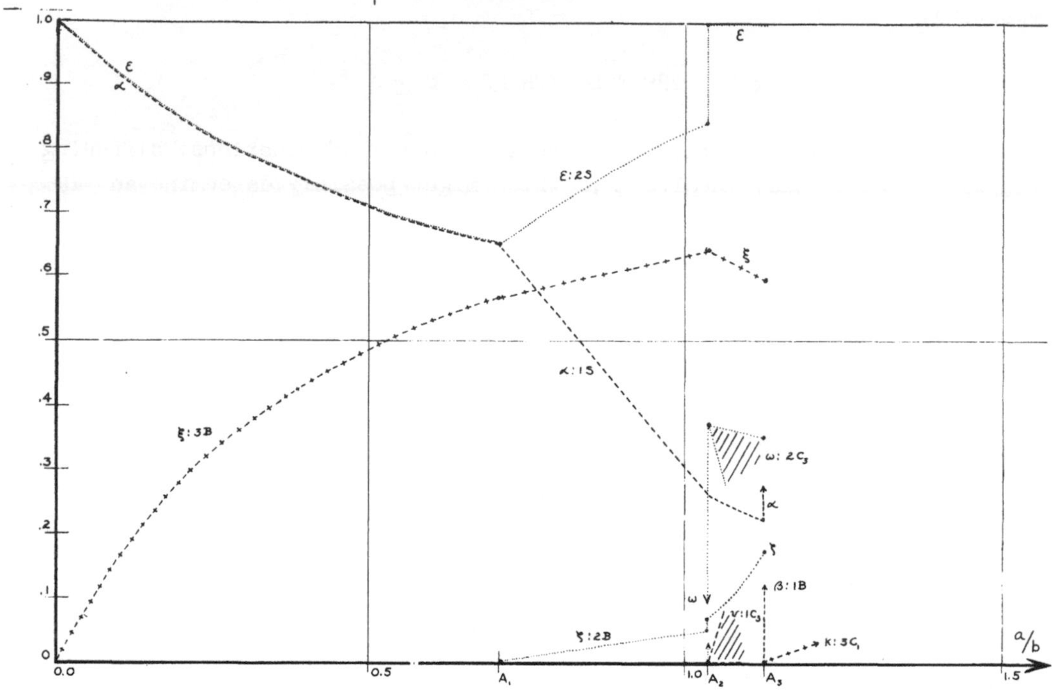

Figure 1A

Equilibrium Point Behavior Coefficients

-----	Player 1	kS:	Player k sandbags
. . . . .	Player 2	kB:	Player k bluffs
+-+-+	Player 3	kC_h:	Player k calls Player h's bluff

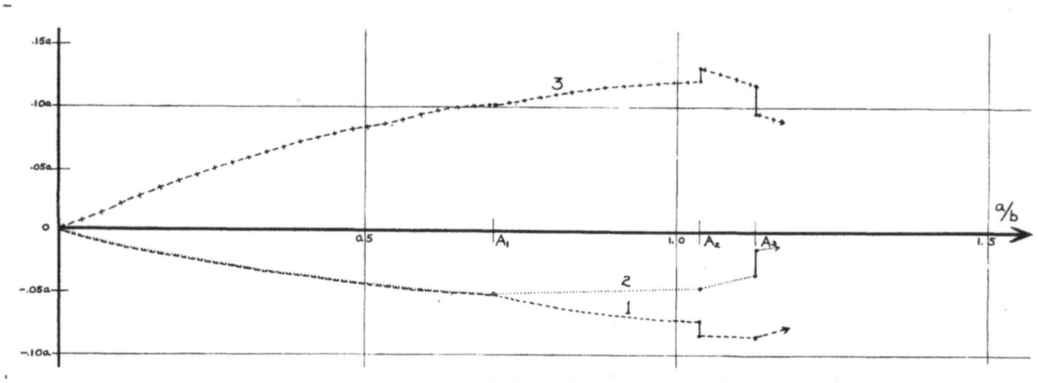

Figure 1B

Value of the Game

Since the EP is unique for each a/b, the value of the game is well-defined throughout R_I and R_{II}. In R_I it is the triple

$$V = \left\langle -\frac{a^2}{8(a+b)}, -\frac{a^2}{8(a+b)}, \frac{a^2}{4(a+b)} \right\rangle .$$

In R_{II} the results are again best given graphically (Figures 1B and 2) and numerically. We have:

$$V = \langle -.0518a, -.0518a, .1036a \rangle \quad \text{at} \quad a/b = A_1 \ ;$$

$$V = \langle -.0735a, -.0479a, .1214a \rangle \quad \text{at} \quad a/b = 1 \ .$$

The connection at A_1 between the two cases is of course continuous.

Thus, Player 3 enjoys a definite advantage. This may be ascribed to his ability to cause a pass-out.[15] He bluffs (ξ) an increasing proportion of the time as the relative size of the ante increases. His opponents meet this strategem by the manoeuvre known as "trapping" or "sandbagging" ($\bar{\alpha}$, $\bar{\varepsilon}$): passing with a high card on the first round, then calling on the second. For the small ante case (a/b in R_I), equilibrium is maintained only if the first two players follow the same strategy ($\alpha = \varepsilon$), and receive the same (negative) payoff. But for larger ante (a/b in R_{II}) it becomes possible for Player 2 also to bluff (ζ) a small fraction of the time, and his fortunes take a turn for the better. Player 1 sharply increases his sandbagging activity ($\bar{\alpha}$) (while Player 2 diminshes his), but this appears to be a defensive tactic only, for his position continues to worsen as a/b increases.

§10. EXTENSION OF THE SOLUTION

The solutions of system (II) can be continued through the range R_{II},

$$1 < a/b < A_2 = 1.0376 \ \ldots \text{[16]}$$

without violating (C). If a/b exceeds A_2 then the value of Δ_ω becomes positive. It is easily verified that we still have an EP in this extension, although uniqueness and the existence of a well-defined value are no longer assured.

---

[15] This is analogous to the advantage in von Neumann's variant "C" (op. cit., pp. 211-218) which accrues to the first player. For that player also has the power to stop the play without forfeiting his basic investment ("b", loc. cit.) and without risking an additional sum ("a - b", loc. cit.). But "initiative," in the sense of having the first move, is a distinct handicap in our game.

[16] A_2 is the positive root of $162A^4+135A^3-144A^2-150A-28$.

At $a/b = A_2$ a qualitatively new phenomenon occurs. System (II) gives the EP:

$$\alpha = .2656,\ \varepsilon = .8442,\ \zeta = .0491,\ \xi = .6423,\ \omega = 0\ .$$

But this is just one extreme of a one-parameter family of EP's, the other extreme of which is the EP:

$$\alpha = .2656,\ \varepsilon = 1,\ \zeta = .0623,\ \xi = .6423,\ \omega = .3720\ .$$

The coefficients ε, ζ, and ω increase together while the others remain fixed. Since just one player's behavior is variable, the family of EP's forms an <u>interchange system</u>: every triple of strategies selected individually from the EP's is itself an EP.

The value of the game at $a/b = A_2$ is not well-defined. Player 3 gains at the expense of Player 1 as ε, ζ, ω increase, while Player 2's share remains (necessarily) constant. The numerical range of values:

$$V = \langle -.0755a,\ -.0475a,\ .1230a \rangle \quad \text{at } a/b = A_2,\ \omega = 0\ ;$$

$$V = \langle -.0826a,\ -.0475a,\ .1301a \rangle \quad \text{at } a/b = A_2,\ \varepsilon = 1\ .$$

Beyond A_2 another new effect appears. If we set $\varepsilon = 1$ and $\Delta_\omega = 0$, we find that $\Delta_\upsilon = 0$ also. This suggests the system:

$$\text{(III)} \qquad \begin{cases} \Delta_\alpha = 0, & 0 \leq \alpha \leq 1 \\ \varepsilon = 1, & \Delta_\varepsilon \geq 0 \\ \Delta_\zeta = 0, & 0 \leq \zeta \leq 1 \\ \Delta_\xi = 0, & 0 \leq \xi \leq 1 \\ \Delta_\upsilon = 0, & 0 \leq \upsilon \leq 1 \\ \Delta_\omega = 0, & 0 \leq \omega \leq 1\ . \end{cases}$$

Solving gives unique values for α, ζ, and ξ, and for the product $\bar{\upsilon}\bar{\omega}$, and puts a restriction on υ,

$$\upsilon \leq \upsilon_{max} \approx 110\ (a/b - A_2)\ .$$

With these values (C) is satisfied throughout the range:

$$R_{III} \qquad A_2 < a/b < A_3 = 1.1262\ \ldots\ .^{17}$$

In R_{III} the EP's are not unique, as just noted, nor do they form an interchange system, since different players control υ and ω. But it is a curious fact that the game does indeed possess a well-defined value (if we

---

[17] A_3 is the largest root of $12A^3 - 14A^2 - 3A + 4$. Δ_β becomes positive if a/b exceeds A_3.

assume that there are no EP's undiscovered). Thus the two properties — existence of a value and interchangeability of equilibrium strategies — which are found in the solution of all two-person, zero-sum games, are quite independent of one another in our three-person Poker.

The situation at $a/b = A_3$ rounds out the picture, for here we have discovered a two-parameter family of EP's. The interchange systems are indexed by one parameter, while the payoffs depend only on the other parameter. The new coefficient to enter at A_3 is β, Player 1's bluff; whereas κ, Player 3's countermeasure to the bluff, comes in immediately thereafter. The value to Player 1 also finally starts to improve for $a/b > A_3$.[18]

§11. THE COALITION GAME

For the sake of making a comparison of our solution with the solutions as defined by von Neumann and Morgenstern,[19] we now calculate the

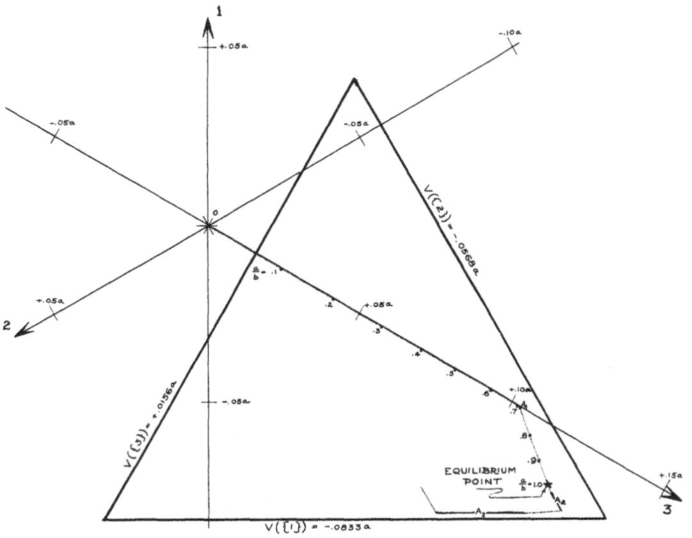

Figure 2
Characteristic Triangle of Coalition Game with $a = b$
(with values of the non-cooperative game for $0 < a/b \leq A_3$)

[18]It is easily verified that for $a \geq 4b$ the value of the game is zero for all players, since each can guarantee himself that amount by betting on all occasions, regardless of his hand.

[19]For the actual construction of the solutions and their interpretation, see the discussion op. cit., particularly pp. 282-290.

"characteristic function" of the game for the case $a = b$ That is, for each set of players (a "coalition") we determine the maximum they can obtain for themselves as a group, irrespective of the actions of the remaining players. Essentially, therefore, we determine the values of three different two-person, zero-sum games. We do this here by exhibiting the optimal strategies.

We cannot expect to be able to represent all the strategies available to a coalition by means of our one-person behavior coefficients.[20] The correlated randomization which may be required is illustrated in the optimal strategy for $\{1, 3\}$ as given below. For the other coalitions, the one-person behavior coefficients happen to be sufficient to describe optimal play.

The optimal strategies for the several coalitions are as follows:

$\{1\}$ $\alpha = 2/3$.	$\{2, 3\}$ $\xi = 2/3$.
$\{2\}$ $\varepsilon = 7/11$, $\zeta = 3/11$.	$\{1, 3\}$ $\begin{cases} \alpha = 0, \ \xi = 3/16 & \text{with prob. } 8/11; \\ \alpha = 1, \ \xi = 0 & \text{with prob. } 3/11. \end{cases}$
$\{3\}$ $\mu = 1/4$, $0 \leq \xi \leq 2/3$.	$\{1, 2\}$ $\alpha = 3/4$, $\varepsilon = 0$.

When not otherwise specified, always bet on High, pass on Low. The optimal strategy is not unique for $\{3\}$, but is unique for all other proper coalitions.

The characteristic function:

$$
\begin{aligned}
v(\{1\}) &= -v(\{2, 3\}) = -a/12 = -.0833a \\
v(\{2\}) &= -v(\{1, 3\}) = -5a/88 = -.0568a \\
v(\{3\}) &= -v(\{1, 2\}) = a/64 = .0156a \\
v(\{1, 2, 3\}) &= v(0) = 0 .
\end{aligned}
$$

Thus, Player 3 has a positive expectation even when the other two are allied against him.

The solutions as defined by von Neumann and Morgenstern are sets of triples lying in the triangle bounded by these values (see Figure 2). The equilibrium point is contained in precisely two of these solutions, both of them of the "discriminatory" type. From Figure 2 it is obvious that the third player has by far the most to fear from collusion between his opponents.

J. Nash
L. S. Shapley

Princeton University

[20]The information available to a two-player coalition is not monotone increasing When making its second move, the coalition has "forgotten" the contents of the hand of its first member.

ISOMORPHISM OF GAMES, AND STRATEGIC EQUIVALENCE

J. C. C. McKinsey[1]

We shall be concerned here with the notion of strategic equivalence in the theory of games.[2] The intuitive notion of strategic equivalence is rather vague; but it nevertheless happens to be sufficiently sharp, that one can specify a precise mathematical condition A, which is intuitively recognized to be necessary for strategic equivalence, and a precise mathematical condition B, which is intuitively recognized to be sufficient. It turns out, however, that B is a consequence of A: so, in actuality, we can say that A (or B) is a necessary and sufficient condition for strategic equivalence. This paper will be devoted mainly to a detailed proof that A implies B.

§1. INTRODUCTION

We shall on occasion avail ourselves of some familiar set-theoretical notation. In particular, if S and T represent sets, then by $S \cap T$, $S \cup T$, and $S - T$, we shall mean, respectively, the intersection, union, and difference of S and T. If x denotes an individual, and S a set, then we shall write $x \in S$ to indicate that x is a member of S. If S and T denote sets, then we shall write $S \subseteq T$ to indicate that S is a subset of T. We shall denote the empty set by Λ. If $a_1, a_2, \ldots, a_n$ denote any kind of things, then by $\{a_1, a_2, \ldots, a_n\}$ we shall mean the set whose only members are $a_1, a_2, \ldots, a_n$, and by $\langle a_1, a_2, \ldots, a_n \rangle$ we shall mean the ordered n-tuple whose first member is a_1, whose second member is a_2, etc.

DEFINITION 1. By a <u>game</u> we shall mean an n-person zero-sum game, with $n > 2$. Thus a game is a real-valued

[1] Received January 14, 1949 by the ANNALS OF MATHEMATICS and accepted for publication; transferred by mutual consent to ANNALS OF MATHEMATICS STUDY No. 24.

[2] The theory of games of strategy has been developed in a systematic way in the treatise: "Theory of Games and Economic Behavior," by J. von Neumann and O. Morgenstern, Princeton, 1947.

function[3] v, which is defined for every subset of the set $I_n = \{1, 2, \ldots, n\}$ of the first n positive integers, and satisfies the following conditions:

(i) $v(\Lambda) = 0$;

(ii) for every subset S of I_n, $v(I_n - S) = -v(S)$;

(iii) if S and T are mutually exclusive subsets of I_n, the $v(S \cup T) \geq v(S) + v(T)$.

The intuitive meaning of a game v is this: if S is any subset of the players I_n, then $v(S)$ is the exact amount the members of S can expect to be paid (or to have to pay, of course, if $v(S)$ is negative) if they form a coalition, and the remaining players form a coalition against them, and everybody behaves in the most intelligent possible way.

DEFINITION 2. By an <u>imputation</u> for a game v, we mean an ordered n-tuple $\langle x_1, x_2, \ldots, x_n \rangle$ of real numbers which satisfy the conditions:

(i) $x_1 + x_2 + \ldots + x_n = 0$

(ii) $v(\{i\}) \leq x_i$, for $i = 1, 2, \ldots, n$.

If v is a game, then we shall denote the set of all imputations for v by $\mathfrak{A}(v)$.

The intuitive meaning of an imputation is that it represents a possible way of dividing money among the players at the end of a play of the game. Condition (i) of Definition 2 expresses the fact that the game is zero-sum: no outside person or agency contributes money to the players (or takes money away from them) after the play. Condition (ii) makes the distribution one which would not be immediately rejected by any one of the players: for if we had $x_i < v(\{i\})$, then player i would be unwilling to take $\langle x_1, x_2, \ldots, x_n \rangle$ into account as a possible distribution of money — since he could obtain more than this distribution offers him even in the worst possible case, when all the other players formed a coalition against him.

---

[3]In a general theory of games, it would be necessary to define a game in a way which would be more complicated from the point of view of set theory. Thus we should have to take account of the players, the nature of the moves available to them, the amounts to be paid them according to the outcome of the play, and the like, and then define a game as an ordered set having these several entities as components. In the present discussion, however, we shall be concerned only with what von Neumann and Morgenstern call "characteristic functions" (op. cit., pp. 238-245); hence it is terminologically convenient to use the word "game" as synonomous with "characteristic function."

DEFINITION 3. We say that an imputation $\langle x_1, x_2, \ldots, x_n\rangle$ of $\mathfrak{A}(v)$ dominates an imputation $\langle y_1, y_2, \ldots, y_n\rangle$ of $\mathfrak{A}(v)$ with respect to a subset S of I_n, if the following conditions are satisfied:

(i) $S \neq \Lambda$;

(ii) for every i in S, $x_i > y_i$;

(iii) $\sum_{i \in S} x_i \leq v(S)$.

To say that an imputation $\langle x_1, x_2, \ldots, x_n\rangle$ dominates an imputation $\langle y_1, y_2, \ldots, y_n\rangle$ with respect to a set S means intuitively that every member of S individually prefers $\langle x_1, x_2, \ldots, x_n\rangle$ to $\langle y_1, y_2, \ldots, y_n\rangle$ (since it pays him more); and that, at the same time, the amount offered the members of S collectively by $\langle x_1, x_2, \ldots, x_n\rangle$ is not unrealistically large: i.e., they recognize, if they are sufficiently intelligent, that it is possible for them actually to obtain this amount, by simply combining together into a coalition.

In the case of each of the terms heretofore introduced, we have had, corresponding to some intuitive idea, a clear and precise mathematical notion. Thus it has been natural to follow the practice of formulating the mathematical notion in a numbered definition, and then subsequently giving an explanation of it from the point of view of the intended intuitive interpretation. With regard to the term to be introduced next, however, the situation is different. For this term we have not immediately available a formulation which would clearly be an exact mathematical representation of our intuitive notion; therefore, in this case, we shall give merely the intuitive explanation, but no formal numbered definition.

By saying that two games are strategically equivalent, we mean that the relative positions of the players, so far as regards their bargaining power, and their tendencies to form (or refrain from forming) coalitions, are the same in the two games. Thus, to say that two games, v and $\bar{v}$, are strategically equivalent means that there exists a one-to-one correspondence between $\mathfrak{A}(v)$ and $\mathfrak{A}(\bar{v})$ such that for every subset S of the set of players, if the member α of $\mathfrak{A}(v)$ corresponds to the member $\bar{\alpha}$ of $\mathfrak{A}(\bar{v})$ and if β similarly corresponds to $\bar{\beta}$, then the tendency of S to choose $\bar{\alpha}$ instead of $\bar{\beta}$ is proportional to the tendency of S to choose α instead of β.

The vagueness of the above explanation of strategic equivalence lies in the phrase "tendency of S to choose." But we see that, when two games are strategically equivalent, then there is a one-to-one correspondence between the sets of their imputations, and this correspondence at least preserves the relation of dominance. This consideration leads us to make the following definition.

DEFINITION 4. A game v is said to be <u>isomorphic</u> to a game $\overline{v}$ with respect to the binary relation $\cong$, if the following conditions hold:

(i) $\cong$ establishes a one-to-one correspondence between the members of $\mathfrak{A}(v)$ and $\mathfrak{A}(\overline{v})$ — i.e., between the sets of imputations of the two games;

(ii) if α and β are any members of $\mathfrak{A}(v)$, and $\overline{\alpha}$ and $\overline{\beta}$ the corresponding members of $\mathfrak{A}(\overline{v})$ — so that $\alpha \cong \overline{\alpha}$ and $\beta \cong \overline{\beta}$ — and if S is any subset of I_n, then α dominates β with respect to S if and only if $\overline{\alpha}$ dominates $\overline{\beta}$ with respect to S.

From the remarks just preceding the above definition, we see that the following principle holds of our intuitive notion of strategic equivalence.

PRINCIPLE 1. If two games are strategically equivalent, then they are isomorphic.

We now introduce another notion which will turn out to be important.

DEFINITION 5. A game v is said to be <u>S-equivalent</u>[4] to a game $\overline{v}$ if there exists a positive constant k, and n constants $a_1, a_2, \ldots, a_n$ whose sum is zero, such that, for every subset S of I_n,

$$\overline{v}(S) = k \cdot v(S) + \sum_{i \in S} a_i .$$

[4] von Neumann and Morgenstern (op. cit., pp. 245-246) use the term "strategic equivalence" for what we here call "S-equivalence" — or rather, properly speaking, for what we might call "S-equivalence with $k = 1$." Their definition is an informal one, however, and it has seemed to us preferable from the point of view of exposition to use the term "strategic equivalence" in its general intuitive meaning.

REMARK. It follows immediately from the definition that the relation of S-equivalence is reflexive, symmetric, and transitive; the same is of course also true of the relation of isomorphism.

We can interpret the k in Definition 5 as meaning a change in the monetary unit. The constants $a_1, a_2, \ldots, a_n$ can be interpreted as fixed payments which are made to the various players at the end of each play — regardless what coalitions have been formed. It is intuitively clear that no essential change would be introduced, if we were to suppose that these fixed payments are made at the start of the play, instead of at the end[5]; thus the constants $a_1, \ldots, a_n$ have no effect on the tendencies to form coalitions. And of course, no change in the tendencies to form coalitions is introduced by a change in the monetary unit. Hence we conclude that the following principle holds of our intuitive notion of strategic equivalence.

PRINCIPLE 2. If two games are S-equivalent then they are strategically equivalent.

REMARK. We have thus shown by intuitive arguments that isomorphism is a necessary condition for strategic equivalence, and that S-equivalence is a sufficient condition; but it does not seem to be intuitively evident that isomorphism is a sufficient condition, nor that S-equivalence is a necessary condition. We shall clear up this question, however, by showing that games are S-equivalent if and only if they are isomorphic. Since von Neumann and Morgenstern have given a formal (non-intuitive)[6] proof that S-equivalence implies isomorphism, it will be necessary only to establish the converse implication.

2.§ ESSENTIAL AND INESSENTIAL GAMES

In this section we shall simplify our task by the elimination of some rather trivial cases.

DEFINITION 6. An n-person game v is said to be in <u>reduced form</u> if $v(\{1\}) = \ldots = v(\{n\}) = c$, where either $c = -1$ or $c = 0$. If $c = 0$, the game is called <u>inessential</u>. If $c = -1$, it is called <u>essential</u>.

---

[5] This argument is taken from von Neumann and Morgenstern (op. cit., p. 246).

[6] Op. cit., p. 281.

LEMMA 1. Every game is S-equivalent to a game in reduced form.

PROOF. Let v be any n-person game. If

$$\sum_{i=1}^{n} v(\{i\}) = 0 ,$$

then, by taking $k = 1$ and, for $i = 1, \ldots, n$, $a_i = -v(\{i\})$, we see that v is S-equivalent to an inessential game in reduced form. If

$$\sum_{i=1}^{n} v(\{i\}) \neq 0 ,$$

then, by taking

$$k = -n/\sum_{i=1}^{n} v(\{i\})$$

and, for $j = 1, \ldots, n$,

$$a_j = -1 + n \cdot v(\{j\})/\sum_{i=1}^{n} v(\{i\}) ,$$

we see that v is S-equivalent to an essential game in reduced form.

The following has been proved by von Neumann and Morgenstern.[7]

LEMMA 2. If v is an inessential game in reduced form, then, for every subset S of I_n, $v(S) = 0$. If v is an essential game in reduced form, and if S is an r-element subset of I_n, then $-r \leq v(S) \leq n - r$.

REMARK. From Lemma 1 we see that, in order to prove that isomorphism implies S-equivalence in general, it suffices to show this for games in reduced form. For if v is isomorphic to u, let v' and u' be games in reduced form which are respectively S-equivalent to u and v. Then u' is isomorphic to v', by the transitivity of isomorphism, together with the fact that S-equivalence implies isomorphism, and hence u' is S-equivalent to v'. Hence, by the transitivity of S-equivalence, we conclude that u is S-equivalent to v, as was to be shown.

From Definition 6, moreover, it follows that the set of imputations for an inessential game in reduced form consists of the single imputation $\langle 0, 0, \ldots, 0\rangle$, while there are always infinitely many

---

[7] Op. cit., p. 249.

imputations for an essential game in reduced form. Thus an essential game cannot be isomorphic (and a fortiori cannot be S-equivalent) to an inessential game. Since, by Lemma 2, there is only one inessential game in reduced form, we see that in order to prove our theorem it suffices to show that every pair of essential games in reduced form which are isomorphic are also S-equivalent. From the fact that the relation of S-equivalence is reflexive, finally, we conclude that it suffices to show that every pair of essential games in reduced form which are isomorphic are also identical.

§3. PROOF THAT ISOMORPHISM IMPLIES S-EQUIVALENCE

Proofs of theorems very similar to the following lemma can be found in the literature.[8]

LEMMA 3. Let f be a bounded function of one real variable which is defined over the interval

$$-1 \leq x \leq 1 ;$$

and suppose that, whenever x, y, and $x + y$ are all in this interval, we have

$$f(x + y) = f(x) + f(y) .$$

Then there exists a number k such that, for every x in the given interval,

$$f(x) = k \cdot x .$$

LEMMA 4. Let n be a positive integer greater than 2, and denote the interval

$$-1 \leq x \leq n - 1$$

by S_n. Let $f_1, \ldots, f_n$ be bounded functions, each of one real variable, which are defined over S_n and satisfy the conditions:

(A) $f_i(-1) = -1$, for $i = 1, \ldots, n$;

(B) if $x_1, \ldots, x_n$ are in S_n and $x_1 + \ldots + x_n = 0$, then $f_1(x_1) + \ldots + f_n(x_n) = 0$.

[8] See, for example, J. G. Darboux, "Sur la composition des forces en statique," Bulletin des Sciences Mathématiques, Vol. 9 (1875), p. 281.

Then, for $i = 1, \ldots, n$, and for any x in S_n, we have

$$f_i(x) = x \ .$$

PROOF.[9] We first show that all the functions $f_1, \ldots, f_n$ are identical. We notice that whenever $x \in S_n$ then $(1 - x)/(n - 2) \in S_n$. Hence, for any x in S_n we have, from (B),

$$f_1(-1) + f_2(x) + f_3[(1-x)/(n-2)] + \ldots + f_n[(1-x)/(n-2)] = 0$$

and also

$$f_1(x) + f_2(-1) + f_3[(1-x)/(n-2)] + \ldots + f_n[(1-x)/(n-2)] = 0 \ .$$

Subtracting the second of these equations from the first, and making use of (A), we conclude that $f_1(x) = f_2(x)$. In similar fashion we see that f_i and f_j are identical for all i and j.

To simplify the notation, we set $f_i = f$, for $i = 1, \ldots, n$. We wish to show that, for every x in S_n, $f(x) = x$.

Denote the interval

$$-1 \leq x \leq 1$$

by S. We shall show first that the conclusion of our lemma holds for every x in S. First, since $0 \in S_n$, we have

(1) $$f(0) = 0 \ .$$

Now if x is any point of S, then $-x$ is a point of S, and of course $x + -x = 0$; thus, making use of (1), we see that, for every x in S,

(2) $$f(-x) = -f(x) \ .$$

Now from (1) and (2) and condition (B) of the hypothesis, we see that, if x, y, and $x + y$ are all in S, then

$$f(x + y) = f(x) + f(y) \ .$$

Thus the hypothesis of Lemma 3 is satisfied, so we conclude that there exists a k such that

$$f(x) = k \cdot x, \text{ for every } x \text{ in } S \ .$$

If we replace x by -1 in the last equation we have, in view of (A)

$$k = -[k \cdot -1] = -f(-1) = 1 \ .$$

Hence

(3) $$f(x) = x, \text{ for every } x \text{ in } S \ .$$

---

[9] I am grateful to the referee for suggesting some simplifications in the proof of this lemma, as well as in the proof of Lemma 6.

Now suppose that x is in S_n but not in S. Then

$$1 < x \leq n - 1 ;$$

consequently

$$- 1/(n-1) > - x/(n-1) \geq - 1 ;$$

and, *a fortiori*,

$$(4) \qquad - 1 \leq - x/(n-1) \leq 1 .$$

Suppose now that we set

$$(5) \qquad x_1 = x, \; x_i = - x/(n-1) \quad \text{for} \quad i = 2, \ldots, n .$$

By hypothesis we have $x_1 \in S_n$, and by (4) we see that $x_i \in S_n$ for $i > 1$. Moreover, it is clear that $x_1 + \ldots + x_n = 0$. Hence we have

$$(6) \qquad f(x_1) + \ldots + f(x_n) = 0 .$$

Since, by (4), $x_i \in S$ for $i > 1$, we conclude by the first part of the proof that $f(x_i) = x_i$ for $i > 1$. Thus from (6) we have

$$f(x_1) + x_2 + \ldots + x_n = 0 ,$$

and hence, from (5)

$$f(x) = x ,$$

as was to be shown.

LEMMA 5. Let v and $\overline{v}$ be essential games in reduced form which are isomorphic under the relation $\cong$, and suppose that $\langle x_1, x_2, \ldots, x_n \rangle \cong \langle \overline{x}_1, \overline{x}_2, \ldots, \overline{x}_n \rangle$. Then, for $i = 1, \ldots, n$, a necessary and sufficient condition that $x_i = -1$ is that $\overline{x}_i = - 1$.

PROOF. It clearly suffices to prove the lemma for the case $i = 1$. By symmetry moreover, it is evident that it suffices to show that, if $x_1 = - 1$, then $\overline{x}_1 = -1$.

Suppose then, if possible, that $x_1 = - 1$ and $\overline{x}_1 \neq - 1$. If we put $\varepsilon = (\overline{x}_1 + 1)/(n-1)$, and set

$$y_1 = - 1$$

$$y_i = \overline{x}_i + \varepsilon \quad \text{for} \quad i = 2, \ldots, n ,$$

then it is easily seen that the sequence $\langle y_1, \ldots, y_n \rangle$ is an imputation which dominates $\langle \overline{x}_1, \ldots, \overline{x}_n \rangle$ with respect to the set $\{2, \ldots, n\}$. By the isomorphism we conclude that there is an imputation $\langle z_1, \ldots, z_n \rangle$ which dominates $\langle x_1, \ldots, x_n \rangle$ with respect to $\{2, \ldots, n\}$. Then we have

$z_i > x_i$ for $i = 2, \ldots, n$, and hence $z_1 < x_1 = -1$. Since this is impossible in view of the fact that $\langle z_1, \ldots, z_n \rangle$ is an imputation of a game in reduced form, our lemma follows.

DEFINITION 7. If $\alpha = \langle a_1, \ldots, a_n \rangle$ and $\beta = \langle b_1, \ldots, b_n \rangle$ are imputations, then by $L(\alpha, \beta)$ we mean the set of integers i such that $a_i < b_i$. By $G(\alpha, \beta)$ we mean the set of integers i such that $a_i > b_i$. By $E(\alpha, \beta)$ we mean the set of integers i such that $a_i = b_i$.

LEMMA 6. Let v and $\overline{v}$ be essential games in reduced form, which are isomorphic under the relation $\cong$, and suppose that $\alpha \cong \overline{\alpha}$ and $\beta \cong \overline{\beta}$. Then

$$G(\alpha, \beta) = G(\overline{\alpha}, \overline{\beta})$$
$$L(\alpha, \beta) = L(\overline{\alpha}, \overline{\beta})$$
$$E(\alpha, \beta) = E(\overline{\alpha}, \overline{\beta}) .$$

PROOF. It suffices to show that, whenever the hypothesis is satisfied, we have $G(\alpha, \beta) \subseteq G(\overline{\alpha}, \overline{\beta})$. For then we can conclude from symmetry that $G(\overline{\alpha}, \overline{\beta}) \subseteq G(\alpha, \beta)$, so that $G(\alpha, \beta) = G(\overline{\alpha}, \overline{\beta})$. Interchanging α and β we can conclude also that $G(\beta, \alpha) = G(\overline{\beta}, \overline{\alpha})$, and hence that $L(\alpha, \beta) = L(\overline{\alpha}, \overline{\beta})$. Finally, $E(\alpha, \beta) = E(\overline{\alpha}, \overline{\beta})$ follows from the fact that two integers i and j satisfy one and only one of the relations: $i < j$, $i = j$, $i > j$.

The proof that the hypothesis implies $G(\alpha, \beta) \subseteq G(\overline{\alpha}, \overline{\beta})$ falls into three parts: I, II and III.

Part I. In this part we shall show that $G(\alpha, \beta) \subseteq G(\overline{\alpha}, \overline{\beta})$ is true whenever the hypothesis is satisfied and $\alpha = \langle a_1, \ldots, a_n \rangle$ and $\beta = \langle b_1, \ldots, b_n \rangle$ are such that: (A) $E(\alpha, \beta) = \Lambda$; (B) for every i in $L(\alpha, \beta)$, $a_i \neq -1$. We shall do this by an induction on the number of elements in $L(\alpha, \beta)$.

If $L(\alpha, \beta)$ contains just one element, say i_0, then, since by hypothesis $E(\alpha, \beta) = \Lambda$, we see that $G(\alpha, \beta) = I_n - \{i_0\}$. Thus α dominates β with respect to the set $I_n - \{i_0\}$. By the isomorphism, we conclude that $\overline{\alpha}$ dominates $\overline{\beta}$ with respect to this set. Thus, if we set $\overline{\alpha} = \langle \overline{a}_1, \ldots, \overline{a}_n \rangle$ and $\overline{\beta} = \langle \overline{b}_1, \ldots, \overline{b}_n \rangle$, we have $\overline{a}_i > \overline{b}_i$ for $i \neq i_0$. Hence $G(\alpha, \beta) \subseteq G(\overline{\alpha}, \overline{\beta})$, as was to be shown.

Now suppose that $G(\alpha, \beta) \subseteq G(\overline{\alpha}, \overline{\beta})$ is true for all α and β which satisfy conditions (A) and (B) and such that $L(\alpha, \beta)$ contains less than k elements. Let α and β be imputations satisfying (A) and (B) and such that $L(\alpha, \beta)$ contains exactly k elements. Without loss of generality we can suppose that

$$-1 < a_i < b_i \quad \text{for } i = 1, \ldots, k,$$
$$a_i > b_i \quad \text{for } i = k+1, \ldots, n.$$

By an algebraic argument of an elementary sort, it can be shown that there exists an imputation $\gamma = \langle c_1, \ldots, c_n \rangle$ which satisfies the conditions:

$$(7) \quad \begin{cases} -1 < c_1 < a_1 < b_1 \\ a_2 < b_2 < c_2 \\ a_i < c_i < b_i, \quad \text{for } i = 3, \ldots, k \\ b_i < c_i < a_i, \quad \text{for } i = k+1, \ldots, n. \end{cases}$$

Now let $\bar{\alpha} = \langle \bar{a}_1, \ldots, \bar{a}_n \rangle$, $\bar{\beta} = \langle \bar{b}_1, \ldots, \bar{b}_n \rangle$, and $\bar{\gamma} = \langle \bar{c}_1, \ldots, \bar{c}_n \rangle$ be the imputations which correspond, respectively, to α, β and γ. From (7) we observe that $L(\alpha, \gamma)$ and $L(\gamma, \beta)$ contain just $k-1$ elements. Hence by the induction hypothesis we see that $G(\alpha, \gamma) \subseteq G(\bar{\alpha}, \bar{\gamma})$ and that $G(\gamma, \beta) \subseteq G(\bar{\gamma}, \bar{\beta})$. Thus we have, for $i = k+1, \ldots, n$, $\bar{a}_i > \bar{c}_i$ and $\bar{c}_i > \bar{b}_i$, and hence $\bar{a}_i > \bar{b}_i$, so that $G(\alpha, \beta) \subseteq G(\bar{\alpha}, \bar{\beta})$, as was to be shown. This completes this part of the proof.

Part II. In this part we show that condition (B) of Part I can be dropped. That is to say we shall show that $G(\alpha, \beta) \subseteq G(\bar{\alpha}, \bar{\beta})$ is true whenever $\alpha = \langle a_1, \ldots, a_n \rangle$ and $\beta = \langle b_1, \ldots, b_n \rangle$ are such that $E(\alpha, \beta) = \Lambda$. We shall do this by an induction on the number of elements in $G(\alpha, \beta)$.

If $G(\alpha, \beta)$ contains just one element, say i_0, then β dominates α with respect to $I_n - \{i_0\}$. By the isomorphism, $\bar{\beta}$ dominates $\bar{\alpha}$ with respect to this same set, so that we have $\bar{b}_i > \bar{a}_i$ for $i \neq i_0$. Since

$$\sum_{i=1}^{n} \bar{b}_i = \sum_{i=1}^{n} \bar{a}_i = 0,$$

we conclude that $\bar{a}_{i_0} > \bar{b}_{i_0}$, so that $G(\alpha, \beta) \subseteq G(\bar{\alpha}, \bar{\beta})$, as was to be shown.

Now suppose that $G(\alpha, \beta) \subseteq G(\bar{\alpha}, \bar{\beta})$ is true whenever $E(\alpha, \beta) = \Lambda$ and $G(\alpha, \beta)$ contains fewer than k elements. Let α and β be imputations such that $E(\alpha, \beta) = \Lambda$, and such that $G(\alpha, \beta)$ contains exactly k elements. Without loss of generality we can suppose that

$$a_i > b_i, \quad \text{for } i = 1, \ldots, k$$

$$a_i < b_i, \quad \text{for } i = k+1, \ldots, n.$$

By an elementary algebraic argument it can be shown that there exists an imputation $\gamma = \langle c_1, \ldots, c_n \rangle$ which satisfies the conditions

$$\begin{aligned} & a_i > c_i > b_i, \quad \text{for } i = 1, \ldots, k-1 \\ (8) \qquad & c_k > a_k > b_k \\ & a_i < c_i < b_i, \quad \text{for } i = k+1, \ldots, n . \end{aligned}$$

Let $\bar{\gamma} = \langle \bar{c}_1, \ldots, \bar{c}_n \rangle$ be the imputation which corresponds to γ. By (8) we see immediately that $E(\alpha, \gamma) = \Lambda$, and that $G(\alpha, \gamma)$ contains just $k - 1$ elements; hence, by the induction hypothesis, we conclude that $G(\alpha, \gamma) \subseteq G(\bar{\alpha}, \bar{\gamma})$; thus

$$(9) \qquad \bar{a}_i > \bar{c}_i, \quad \text{for } i = 1, \ldots, k-1 .$$

From (8) we see that $c_1 > -1$ for $i = 1, \ldots, n$. Moreover, $E(\gamma, \beta) = \Lambda$. Thus γ and β satisfy conditions (A) and (B) of Part I. We therefore conclude that $G(\gamma, \beta) \subseteq G(\bar{\gamma}, \bar{\beta})$, and hence that

$$(10) \qquad \bar{c}_i > \bar{b}_i, \quad \text{for } i = 1, \ldots, k-1 .$$

From (9) and (10) we obtain

$$(11) \qquad \bar{a}_i > \bar{b}_i, \quad \text{for } i = 1, \ldots, k-1 .$$

By exactly the same method used to derive (11), except that we choose γ in a slightly different way, we show that

$$(12) \qquad \bar{a}_i > \bar{b}_i, \quad \text{for } i = 2, \ldots, k .$$

From (11) and (12) we see that $G(\alpha, \beta) \subseteq G(\bar{\alpha}, \bar{\beta})$ as was to be shown. This completes the second part of the proof.

Part III. In this, the final part of the proof we show that the condition $E(\alpha, \beta) = \Lambda$ can be dropped: that is to say, that $G(\alpha, \beta) \subseteq G(\bar{\alpha}, \bar{\beta})$ is true without any restriction on α and β.

Suppose, then, that α and β are any imputations, and that $\bar{\alpha}$ and $\bar{\beta}$ are the corresponding imputations. Without loss of generality, we can suppose that

$$\begin{aligned} & a_i > b_i, \quad \text{for } i = 1, \ldots, r \\ & a_i < b_i, \quad \text{for } i = r+1, \ldots, s \\ & a_i = b_i, \quad \text{for } i = s+1, \ldots, n . \end{aligned}$$

By an elementary argument we can show that there exists an imputation $\gamma = \langle c_1, \ldots, c_n \rangle$ which satisfies the conditions

$$a_i > c_i > b_i, \quad \text{for } i = 1, \ldots, r$$

(13) $$a_i < c_i < b_i, \quad \text{for } i = r + 1, \ldots, s$$

$$a_i < c_i > b_i, \quad \text{for } i = s + 1, \ldots, n \,.$$

Let $\bar{\gamma} = \langle \bar{c}_1, \ldots, \bar{c}_n \rangle$ be the imputation which corresponds to γ. From (13) we see that $E(\alpha, \gamma) = E(\gamma, \beta) = \Lambda$. Hence, by Part II of the proof, we have

$$\bar{a}_i > \bar{c}_i > \bar{b}_i \quad \text{for } i = 1, \ldots, r \,,$$

and hence $G(\alpha, \beta) \subseteq G(\bar{\alpha}, \bar{\beta})$, which completes the proof of our lemma.

The following theorem contains the principal result of this paper.

THEOREM 1. Two games are isomorphic if and only if they are S-equivalent.

PROOF. As was remarked in §2, it suffices to show that every pair of essential games in reduced form which are isomorphic are also identical.

Suppose then, that v and $\bar{v}$ are two essential games in reduced form which are isomorphic under the relation $\cong$.

We define n functions $g_1, \ldots, g_n$ of n real variables by setting

$$g_i(x_1, \ldots, x_n) = z$$

if and only if $\langle x_1, \ldots, x_n \rangle$ is an imputation of $\mathfrak{A}(v)$ and

$$\langle x_1, \ldots, x_n \rangle \cong \langle y_1, \ldots, y_n \rangle$$

where $y_i = z$. Thus, for any imputation $\langle x_1, \ldots, x_n \rangle$ of $\mathfrak{A}(v)$, and any imputation $\langle y_1, \ldots, y_n \rangle$ of $\mathfrak{A}(\bar{v})$, we have

$$\langle x_1, \ldots, x_n \rangle \cong \langle y_1, \ldots, y_n \rangle$$

if and only if

$$y_i = g_i(x_1, \ldots, x_n), \quad \text{for } i = 1, \ldots, n \,.$$

From Lemma 6 we see that, if $g_1(u_1, \ldots, u_n)$ and $g_1(v_1, \ldots, v_n)$ are defined (i.e., if $\langle u_1, \ldots, u_n \rangle$ and $\langle v_1, \ldots, v_n \rangle$ are imputations), and if $u_1 = v_1$, then $g_1(u_1, \ldots, u_n) = g_1(v_1, \ldots, v_n)$. Thus, for any imputations $\langle x_1, u_2, \ldots, u_n \rangle$ and $\langle x_1, v_2, \ldots, v_n \rangle$, we have

$$g_1(x_1, u_2, \ldots, u_n) = g_1(x_1, v_2, \ldots, v_n) \,.$$

Hence the values of g_1 depend only on the first argument; so there exists an f_1 such that, for all imputations $\langle x_1, x_2, \ldots, x_n \rangle$,

$$g_1(x_1, x_2, \ldots, x_n) = f_1(x_1) .$$

Similarly, for $i = 2, \ldots, n,$ we can set

$$g_i(x_1, \ldots, x_n) = f_i(x_i) .$$

Hence there are n functions $f_1, \ldots, f_n,$ each of one variable, such that for all imputations $\langle x_1, \ldots, x_n \rangle$ and $\langle y_1, \ldots, y_n \rangle$,

$$\langle x_1, \ldots, x_n \rangle \cong \langle y_1, \ldots, y_n \rangle$$

is true if and only if

$$y_i = f_i(x_i), \quad \text{for} \quad i = 1, \ldots, n .$$

By means of Lemma 5 we see that, for $i = 1, \ldots, n,$.

$$f_i(-1) = -1 .$$

Hence, making use of the definition of an imputation, we see that the hypothesis of Lemma 4 is satisfied by the functions $f_1, \ldots, f_n$. Therefore we conclude that, for $i = 1, \ldots, n,$ and for every x,

$$f_i(x) = x .$$

Thus if α and β are imputations, and $\alpha \cong \beta$, then $\alpha = \beta$.

For any subset S of I_n, the value of $v(S)$ can be computed as the maximum of the values of

$$\sum_{i \in S} a_i$$

for $\langle a_1, \ldots, a_n \rangle$ an imputation which dominates some imputation with respect to S. Since the same is true for the value of $\overline{v}(S)$, we conclude from the fact that v and $\overline{v}$ are isomorphic under the identity mapping, that $v(S) = \overline{v}(S)$ is true for all S, as was to be shown.

We conclude with a theorem which sharpens the intuitive notion of strategic equivalence.

THEOREM 2. Principles 1 and 2 of §1 imply that strategic equivalence is the same as S-equivalence.

PROOF. By Theorem 1.

J. C. C. McKinsey

The RAND Corporation

Part II

INFINITE GAMES

OPERATOR TREATMENT OF MINMAX PRINCIPLE

Samuel Karlin[1]

In recent years due to the stimulus provided by the theory of games developed by J. von Neumann[2] much interest has been given to the question of when

$$(*) \quad \min_g \max_f \int_0^1 \int_0^1 K(x, y)df(x)dg(y) = \max_f \min_g \int_0^1 \int_0^1 K(x, y)df(x)dg(y) ,$$

where f and g denote distributions over the unit interval. In the case where $K(x, y)$ is a simple finite valued function, the integral reduces to a matrix. This case has been investigated by von Neumann. Contributions to the continuous case have been given by J. Ville [2, 3] and A. Wald [4]. Furthermore, J. Ville has shown that relation (*) does not hold for every function $K(x, y)$ defined on the unit interval. Of particular interest in applications to problems of tactical games is the case where $K(x, y)$ is bounded and discontinuous only along the diagonal $x = y$. It is the purpose of this paper to analyze the validity of the relationship (*) under very general conditions. The method used is to introduce the operator

$$Tg = \int K(x,y)df(y)$$

and study various conditions placed upon T to insure the truth of (*).

Paragraph 1 shows that if we assume that the operator T is weakly completely continuous and we require an additional condition of separability on the space of distribution, we secure the truth of (*). The results here contain all known cases.

Paragraph 2 applies and relates the methods of paragraph 1 to other results.

§1. DEFINITIONS

If we consider the operator $Tg(y)$ defined on the space of functions of bounded variation (B. V.) mapping into the bounded functions (B) by means of

---

[1]Received November 18, 1948 by the ANNALS OF MATHEMATICS and accepted for publication; transferred by mutual consent to ANNALS OF MATHEMATICS STUDY No. 24.

California Institute of Technology. The author would like to express his gratitude to Professor H. F. Bohnenblust for valuable discussions.

[2]All references are at the end.

$$Tg = \int k(x, y)dg(y)$$

and we use the notation

$$(f, Tg) = \int df(x) \int k(x, y)dg(y) ,$$

then we shall show under general conditions on the operator that

$$\inf_g \sup_f (f, Tg) = \sup_f \inf_g (f, Tg) .$$

The integrals are taken over the unit square and $k(x, y)$ is assumed to be Borel measurable, bounded and measurable for each variable in the other. In all that follows g and f will always specify distributions, i.e., a non-decreasing function $f(x)$ with $f(0) = 0$ $f(1) = 1$ and such that f is continuous to the right.

DEFINITION 1. Let R be the set of f which corresponds to points of x (R consists of pure strategies).

DEFINITION 2. Let P be the set of f which consists of finite convex combinations of elements of R. It is clear that P consists of all distributions consisting of step functions with a finite number of jumps.

DEFINITION 3. A transformation T is said to be weakly completely continuous (w.c.c) if for any sequence of elements g_n there exists a subsequence g_{m_k} and g_0 such that for every f, we have

$$(f, Tg_{m_k}) \longrightarrow (f, Tg_0) .$$

The convergence is not necessarily uniform in f.

In what follows all the definitions and remarks will apply for f in P.

REMARK 1. The range of Tg can be topologized as follows: We define the neighborhood of a point Tg_0 as

$$U_{f_1, \ldots, f_n, \varepsilon} (Tg_0) = [Tg, \{|(f_i, Tg) - (f_i, Tg_0)| \leq \epsilon\} \ i = 1, \ldots, n] .$$

If the transformation T is w.c.c., then the topological space defined is sequentially compact.

For completeness, we give a short proof of the min max theorem for finite matrices. Consider a matrix a_{ik}. Let the column vectors be denoted by P_i $i = 1, 2, \ldots, n$. Let Γ denote the convex set in n-dimensional Euclidean space spanned by the points $P_i (\Gamma = \mathrm{Co}\,(P_1, \ldots, P_n))$. We call the real number λ admissible if there exists an element P in Γ such that $P \leq \lambda$ (natural partial ordering of the real numbers in n-dimensional space, i.e., each component of P is smaller than λ, where P is of the form

$$P = \sum \xi_i P_i \quad \xi_i \geq 0 \quad \sum \xi_i = 1 .$$

Let $\lambda_0 = \inf \lambda$ with λ admissible). A simple compactness argument shows that λ_0 is also admissible. The set Γ and the set R of all points Q in Euclidean n-dimensional space which satisfy the property $Q \leq \lambda_0$ are non-over-lapping. Hence there exists a plane η^0 which separates Γ and R, i.e., if $\eta_1^0, \ldots, \eta_n^0$ denote the normalized coefficients of the plane $(\sum \eta_i^0 = 1)$, then for some value c, we have

$$(\eta^0, \sum \xi_i P_i) = \sum \eta_i^0 \sum a_{ij} \xi_j \geq c \quad \text{and} \quad (\eta^0, Q) \leq c .$$

It follows immediately that $c = \lambda_0$. If we take Q successively to be $(-N, \lambda_0, \ldots, \lambda_0), (\lambda_0, -N, \lambda_0, \ldots) \ldots$ with N sufficiently large, we deduce that $\eta_i^0 \geq 0$. Thus there exists for all $\xi_j \geq 0, \sum \xi_j = 1$, $\sum \eta_i^0 a_{ij} \xi_j \geq \lambda_0$. By the definition of λ_0 there exists a ξ_j^0 such that for all $\eta_i, \eta_i \geq 0$ $\sum \eta_i = 1$ with $\sum \eta_i a_{ij} \xi_j^0 \leq \lambda_0$. This implies $\max_\eta \min_\xi A(\eta, \xi) = \min_\xi \max_\eta A(\eta, \xi)$ and the fundamental Theorem of Games of von Neumann.

Several Lemmas. All the following lemmas are valid also with the inequalities reversed. They shall be applied in both forms.

LEMMA 1. If $f_1, \ldots, f_n$ are a fixed finite sequence of elements and $\Gamma = [g_1, \ldots, g_k]$ denotes the convex set spanned by $g_1, \ldots, g_k$, then if for every g in Γ there exists an f_i with $(f_i, Tg) \leq \lambda$, then there exists a finite collection of ξ_i $i = 1, \ldots, n$, $\xi_i \geq 0$ with

$$\sum_{i=1}^{m} \xi_i = 1 \quad \text{such that} \quad (\sum_{i=1}^{m} \xi_i f_i, Tg) \leq \lambda$$

for every g in Γ.

PROOF. We consider the matrix

$$a_{ij} = (f_i, Tg_j) .$$

Then the hypothesis states that for each

$$\eta_j \geq 0 \quad \sum_{j=1}^{k} \eta_i = 1$$

there exists an i with $\sum a_{ij}\eta_j \leq \lambda$. If we consider the game generated by the matrix A, then for each η there exists a ξ_η with $A(\xi_\eta, \eta) \leq \lambda$. This implies that $\min_\xi A(\xi, \eta) \leq \lambda$ for every η or $\max_\eta \min_\xi A(\xi, \eta) \leq \lambda$. But the finite game has $\max_\eta \min_\xi A(\xi, \eta) = \min_\xi \max_\eta A(\xi, \eta) \leq \lambda$, and hence possesses a solution. Therefore, there exists a ξ_0 with $A(\xi_0, \eta) \leq \lambda$ for all η. This implies the conclusion of the lemma.

LEMMA 2. If $f_1, \ldots, f_n$ is a fixed finite sequence, and if for every g there exists an f_i with $(f_i, Tg) \leq \lambda$, then there exists a

$$\xi_i \geq 0 \quad \sum_{i=1}^{n} \xi_i = 1$$

with $(\sum \xi_i f_i, Tg) \leq \lambda$ for all g.

PROOF. If we consider the mapping of g into n dimensional space by means of

$$\{(f_1, Tg), (f_2, Tg), \ldots, (f_n, Tg)\} ,$$

then we conclude the existence of a countable set of g_k such that for every $\epsilon > 0$ for every g, there exists a g_{k_0} such that

$$|(f_i, Tg) - (f_i, Tg_{k_0})| \leq \epsilon \quad i = 1, \ldots, n$$

we consider the set $\Gamma_m = [g_1, \ldots, g_m]$ which consists of the convex set spanned by $g_1, \ldots, g_m$. Applying Lemma 1, we conclude the existence of a

$$\xi^m (\xi_i^m \geq 0 \quad \sum_{i=1}^{n} \xi_i^m = 1) \quad \text{such that} \quad (\sum_{i=1}^{n} \xi_i^m f_i, Tg) \leq \lambda$$

for every g in Γ_m. We select a limit point of $\xi^m (\xi^m \longrightarrow \xi)$, then for every g_k, we get

$$(\sum_{i=1}^{n} \xi_i f_i, Tg_k) \leq \lambda .$$

Since g_k is weakly dense as explained above, this implies that $(\sum \xi_i f_i, Tg) \leq \lambda$ for all g. This completes the proof.

DEFINITION 4. Let L denote the set of all λ for which there exists a g such that for all f in P

$$\lambda \geq (f, Tg) .$$

If $K(x, y)$ is such that $|K(x, y)| \leq C$, then clearly the set L is non-empty and bounded below. Let $\lambda_0 = \text{Inf } \lambda$ for λ in L.

LEMMA 3. If T is w.c.c., then λ_0 is in L.

PROOF. Consider a sequence $\lambda_n \longrightarrow \lambda_0$ with λ_n in L, then there exists a sequence of g_n corresponding to λ_n with $\lambda_n \geq (f, Tg_n)$ for all f in P. The hypothesis on T gives the existence of a g_0 with $(f, Tg_{n_k}) \longrightarrow (f, Tg_0)$ for every f in P. The conclusion is now evident.

LEMMA 4. Given an $\epsilon > 0$, if for every f in P there exists a g_f such that $\lambda_0 - \epsilon > (f, Tg_f)$, then for every finite number $f_1, \ldots, f_n$ in P there exists a g with $\lambda_0 - \epsilon > (f_i, Tg)$ for $i = 1, \ldots, n$.

PROOF. If we suppose the contrary, then for every g there exists an i with $(f_i, Tg) \leq \lambda_0 - \epsilon$. Applying Lemma 2 there exists a ξ_i with $\xi_i \geq 0$ $\sum \xi_i = 1$ and $(\sum \xi_i f_i, Tg) \leq \lambda_0 - \epsilon$ for every g. Clearly $\sum \xi_i f_i$ is an element of P and hence there exists by hypothesis an element g such that $(\sum \xi_i f_i, Tg) > \lambda_0 - \epsilon$ which contradicts the above.

DEFINITION 5. We say that T is weakly separable over P if for every $\epsilon > 0$ there exists a sequence of f_i in P such that for every $f \in P$ and g there exists a f_{i_0} depending on f and g with

$$|(f, Tg) - (f_{i_0}, Tg)| \leq \epsilon .$$

LEMMA 5. If λ_0 is defined as in Definition 3 and if T is w.c.c. and weakly separable, then there exists a $f \in P$ with $(\lambda_0 - \epsilon) \leq (f, Tg)$ for all g.

PROOF. If we suppose the contrary, then for every f in P there exists a g such that $\lambda_0 - \epsilon > (f, Tg)$. In virtue of Lemma 4 for any finite number of f_i $i = 1, \ldots, n$, the set of g for which $\lambda_0 - \epsilon > (f_i, Tg)$ $i = 1, \ldots, n$ is non-empty. Let f_i be a weakly dense set (see Definition 5) corresponding to $\frac{\epsilon}{2}$, then for every $f_1, \ldots, f_k$ there exists a g_k such that

$$\lambda_0 - \epsilon > (f_i, Tg_k) \quad \text{for} \quad i = 1, \ldots, k.$$

In virtue of the hypothesis there exists a subsequence g_{k_n} and g_0 such that for every f $(f, Tg_{k_n}) \longrightarrow (f, Tg_0)$, hence $\lambda_0 - \epsilon \geq (f_i, Tg_0)$ for every i. Since f_1 is weakly separable corresponding to $\frac{\epsilon}{2}$ we conclude for every f in P $\lambda_0 - \frac{\epsilon}{2} \geq (f, Tg_0)$. This contradicts the definition of λ_0 and the proof is complete.

THEOREM 1. If T is w.c.c. and weakly separable, then

$$\min_g \sup_f (f, Tg) = \sup_f \min_g (f, Tg) .$$

PROOF. We first remark that for fixed g

$$\sup_f (f, Tg) = \sup_{f \text{ in } P} (f, Tg) .$$

In fact, if $\varphi(x) = \int k(x, y)dg(y)$ we must show that

$$\sup_{\text{all } f} \int \varphi(x)df(x) = \sup_{f=P} \int \varphi(x)df(x) .$$

Let $\int \varphi(x)df(x) = v$ for an f, then if $\varphi(x)$ takes on the value v at x_0, then choose f in P which is a pure strategy at the point x_0. If $\varphi(x)$ does not take on the value v, then there exists values x_1, x_2 where $\varphi(x_1) > v$ and $\varphi(x_2) < v$. Let t be determined so that $t\,\varphi(x_1) + (1 - t)\,\varphi(x_2) = v$. Choose f in P so that f has two jumps, one at x_1 of value t and the other at x_2 of value $(1 - t)$. We proceed now to the proof of Theorem 1. Since λ_0 is in L (Lemma 3), there exists a g_0 such that $\lambda_0 \geq (f, Tg_0)$ for all f in P. Consequently

$$(1) \quad \lambda_0 \geq \sup_{f \in P} (f, Tg_0) = \sup_f (f, Tg_0) \geq \inf_g \sup_f (f, Tg) \geq \sup_f \inf_g (f, Tg) .$$

In virtue of Lemma 5 given an $\epsilon > 0$ there exists a f_0 with $\lambda_0 - \epsilon \leq (f_0, Tg)$ for all g and thus

$$(2) \qquad \lambda_0 - \epsilon \leq \inf_g (f_0, Tg) \leq \sup_f \inf_g (f, Tg) .$$

Combining (1) and (2), since (2) holds for any $\epsilon > 0$, we get $\sup_f (f, Tg_0) = \inf_g \sup_f (f, Tg) = \sup_f \inf_g (f, Tg)$. This proves the Theorem.

Let $\overline{T}f$ denote the operator

$$\overline{T}f = \int k(x, y)df(x) .$$

The operator $\overline{T}$ is contained in the conjugate transformation to T.

THEOREM 2. If in addition to the assumptions of Theorem 1 we assume that $\overline{T}f$ is also w.c.c., then

$$\max_f \min_g (f, Tg) = \min_g \max_f (f, Tg) .$$

PROOF. It is sufficient to show that there exists a f_0 such that $\lambda_0 \leq (f_0, Tg)$ for every g. In virtue of Lemma 5 for every $\epsilon > 0$ there exists an f_ϵ such that $\lambda_0 - \epsilon \leq (f_\epsilon, Tg)$ for all g. Let $\epsilon_n \longrightarrow 0$, then there exists a sequence f_n with $\lambda_0 - \epsilon_n \leq (f_n, Tg) = (\overline{T}f_n, g)$ for every g. In virtue of the hypothesis, we select a subsequence f_{n_k} and f_0 with $(\overline{T}f_{n_k}, g) \longrightarrow (\overline{T}f_0, g)$, and hence $(\overline{T}f_0, g) \geq \lambda_0$ for every g. Q.E.D. We generalize the result of Theorem 1 by removing the hypothesis of weak separability. In place, we have:

THEOREM 3. If the weak topology introduced in Remark 1 is bicompact, then

$$\sup_f \min_g (f, Tg) = \min_g \sup_f (f, Tg) .$$

PROOF. It is sufficient to establish the result of Lemma 5. Indeed, if we suppose the contrary, then for every f in P there exists a g_f such that $\lambda_0 - \epsilon > (f, Tg)$. In virtue of Lemma 4 for any finite number of f_i $i = 1, \ldots, n$ the set of g for which $\lambda_0 - \epsilon > (f_i, Tg)$ $i = 1, \ldots, n$ is non-empty. We well order the elements of P

$$f_1, f_2, \ldots, f_\xi, \ldots$$

let $G(f_1) = \text{closure } [Tg | \lambda_0 - \epsilon > (f_1, Tg)]$, and

$$G(f_{\alpha_1}, \ldots, f_{\alpha_n}) = \text{closure } [Tg | \lambda_0 - \epsilon > (f_{\alpha_i}, Tg) \quad i = 1, \ldots, n] .$$

We define

$$G(f_\xi) = \bigcap G(f_\xi, f_{\alpha_1}, \ldots, f_{\alpha_n})$$

where the intersection is taken over all finite subsets with $\alpha_1, \ldots, \alpha_n$ less than ξ. If $G(f_\xi) = 0$, then by the bicompactness there exists a finite product such that

$$0 = \bigcap_{k=1}^{m} G(f_\xi, f_\alpha k_1, \ldots, f_\alpha k_{n_k}) = G(f_\xi, f_{\alpha_1} 1, \ldots, f_{\alpha_{n_1}} 1, f_{\alpha_1} 2, \ldots, f_{\alpha_{n_2}} 2, \ldots, f_{\alpha_{n_k}} k)$$

which is impossible. Furthermore,

$$G(f_1) \supset G(f_2) \supset \ldots G(f_\xi) \supset G(f_\eta) \supset \ldots$$

with $\eta > \xi$ and each $G(f_\alpha)$ is closed and non-empty. Thus there exists a point Tg_0 in common and hence $\lambda_0 - \epsilon \geq (f, Tg_0)$ for every f. This contradicts the definition of λ_0. Q.E.D. We now weaken further the conditions.

DEFINITION 6. An operator T will be said to be quasi weakly completely continuous if for any sequence g_n given an $\epsilon > 0$, there exists a subsequence g_{n_k} and g_o such that for every f except a finite number at most there exists a $n_o(f, \epsilon)$ such that

(1) $$|(f, Tg_{n_k}) - (f, Tg_o)| \leq \epsilon \text{ for } n_k \geq n_o$$

and f in R. (The g_{n_k} is independent of ϵ.)

DEFINITION 7. An operator T is called densely weakly separable if for any $\epsilon > 0$ there exists a weakly separable sequence f_n in R with the additional property that if you remove any finite number of f_n for which (1) fails to hold, then the remaining sequence remains weakly separable for ϵ.

THEOREM 4. If T is quasi weakly completely continuous and densely weakly separable, then
$$\inf_g \sup_f (f, Tg) = \sup_f \inf_g (f, Tg) .$$

PROOF. Lemma 1 and Lemma 2 remain unchanged. Lemma 3 is not needed since the definition of λ_o implies that for any $\epsilon > 0$ there exists a g_o such that $\lambda_o + \epsilon \geq (f, Tg_o)$ for all f in P. Lemma 4 is unchanged. In order to verify Lemma 5 we have, as in the proof, that for any finite number $f_1, \ldots, f_n$ there exists a g such that

$$\lambda_o - \epsilon > (f_i, Tg) \quad \text{for} \quad i = 1, \ldots, n .$$

Let $\epsilon' = \frac{\epsilon}{4}$, then there exists in virtue of the hypothesis a dense sequence f_n. For every $f_1, \ldots, f_k$ there exists a g_k such that

$$\lambda_o - \epsilon > (f_i, Tg_k) \text{ for } i = 1, \ldots, k .$$

Applying the hypothesis again we get a subsequence g_{n_k} and a g_o such that for $\epsilon' = \frac{\epsilon}{4}$ we have

$$\lambda_o - \tfrac{3}{4}\epsilon \geq (f_i, Tg_o)$$

except for at most a finite number of f_i. The property of the dense sequence f_n implies that

$$\lambda_o - \tfrac{\epsilon}{2} \geq (f, Tg_o)$$

for every f in R and hence f in P. This gives a contradiction. The remainder of the proof is the same.

THEOREM 5. If in addition to the assumptions of Theorem 4 it is supposed that the sequence f_n, guaranteed by Definition 7, is valid for every ϵ, then

$$\min_g \sup_f (f, Tg) = \sup_f \min_g (f, Tg) .$$

PROOF. We must show that Lemma 3 holds; that is, that λ_0 is in L. Let $\lambda_0 + \epsilon_n$ denote a sequence tending to λ_0 with $\epsilon_n > 0$. Let g_n denote the corresponding elements to $\lambda_n = \lambda_0 + \epsilon_n$. Moreover, let f_k denote the dense set guaranteed by the hypothesis. There exist a subsequence g_{n_k} and a g_0 such that, except for a finite number of f in R, for a fixed ϵ_n, $|(f, Tg_{n_k}) - (f, Tg_0)| \leq \epsilon_n$. This implies that except for a finite number of f

$$\lambda_0 + 2\epsilon_{n_k} \geq (f, Tg_0) .$$

But since f_k is densely separable, this holds for every f in R and hence for f in P. But ϵ_n is arbitrary and thus $\lambda_0 \geq (f, Tg_0)$ for every f in P which establishes the assertion.

EXAMPLE 1: Let

$$k(x, y) = \begin{Bmatrix} M(x, y) & x \leq y \\ M'(x, y) & x > y \end{Bmatrix} \quad \text{in I} \quad \begin{matrix} 0 \leq x \leq 1 \\ 0 \leq y \leq 1 \end{matrix}$$

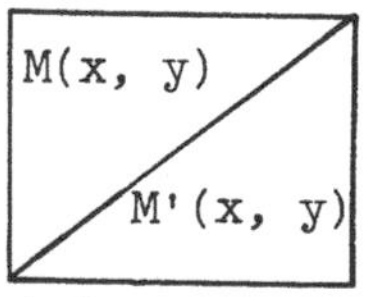

with $M(x, y)$ and $M'(x, y)$ continuous in their respective regions of definitions. We verify that T satisfies the requirements of Definitions 6 and 7. In fact, let g_n denote any sequence and select a subsequence g_e convergent to g_0 at every point of continuity. This is possible by Helly's theorem. Let y_n denote the points of discontinuity of g_0 and let $y_{n_1}, \ldots, y_{n_k}$ denote the finite set of points where the variation of g_0 is greater than $\frac{\epsilon}{4C}$. Then for every x except $y_{n_1}, \ldots, y_{n_k}$, we have that

$$|(x, Tg_e) - (x, Tg_0)| \leq \epsilon \quad \text{for} \quad n_k \geq n_0(x, \epsilon) .$$

Indeed, let x be a point of continuity of g_0. Then

$$\int k(x, y)d(g_e - g_0)(y) = \int_0^{x-\eta} k(x, y)dg_e - g_0(y) + \int_{x-\eta}^{x+\eta} + \int_{x+\eta}^{1} = I_1 + I_2 + I_3$$

$$|I_2| = |\int_{x-\eta}^{x+\eta} k(x,y)dg_e - g_0(y)| \leq C[g_e(x+\eta) - g_e(x-\eta) + g_0(x+\eta) - g_0(x-\eta)],$$

we can choose η so that $|I_2| \leq \epsilon$. For,

$$\lim_{e\to\infty} \int_{x-\eta}^{x+\eta} dg_e = \int_{x-\eta}^{x+\eta} dg_0 = g_0(x+\eta) - g_0(x-\eta)$$

which can be made arbitrarily small since x is a point of continuity. Having chosen η fixed since $k(x, y)$ is continuous in each of integrals I_1 and I_3, we obtain

$$\overline{\lim_{e \to \infty}} \left| \int k(x, y) dg_e - g_0(y) \right| \leq \epsilon .$$

But ϵ is arbitrary and the conclusion is now evident. If x_0 is a point of discontinuity different from $y_{n_1}, \ldots, y_{n_k}$, then the jump of g_0 at x_0 is less than or equal to $\frac{\epsilon}{4C}$. Using the same decomposition it follows that $|I_2| \leq C \cdot \frac{\epsilon}{4C} + 0\ (1)$ for ϵ chosen sufficiently small and hence as before

$$\left| \int k(x, y) d(g_e - g_0) y \right| \leq \epsilon$$

for e chosen sufficiently large. We now verify that Definition 7 is satisfied. Let the dense set in R be all the rational points. We must show that for every element x_0 in R for a given ϵ there exists a rational point f_{x_1} of R depending upon g such that

$$|(f_{x_0}, Tg) - (f_{x_1}, Tg)| < \epsilon$$

or the existence of x_1 such that

$$\left| \int [k(x_0, y) - k(x_1, y)]\, dg(y) \right| < \epsilon .$$

Let η be chosen sufficiently small so that the variation of $g(y)$ from $g(x_0 - \eta)$ till $g(x_0 - 0)$ is less than $\frac{\epsilon}{4C}$ [$g(y - 0)$ shall denote left hand limit], we then choose x_1 to the left of x_0 with $|x_1 - x_0| < \eta$. Put $\epsilon' = \frac{\epsilon}{6}$. It will be clear later that $x_0 = 0$ causes no difficulty. Now,

$$|M(x_0, y) - M(x_1, y)| \leq \epsilon' \quad \text{for} \quad 1 \geq y \geq x_0$$

$$|M'(x_0, y) - M'(x_1, y)| \leq \epsilon' \quad \text{for} \quad 0 \leq y \leq x_0 - \eta ,$$

then

$$\int [k(x_0, y) - k(x_1, y)]\, dg(y)$$

$$= \int_0^{x_0 - \eta} + \int_{x_0 - \eta}^{x_0 - 0} + [g(x_0) - g(x_0 - 0)][M(x_0, x_0) - M(x_1, x_0)]$$

$$+ \int_{x_0}^{1} = I_1 + I_2 + I_3 + I_4 .$$

Clearly $I_1 + I_3 + I_4 \leq 3\epsilon' = \frac{\epsilon}{2}$ and $I_2 \leq 2C \text{ var } g(t) \leq \frac{\epsilon}{2}$ for $x_0 - \eta \leq t \leq x_0 - 0$ or

$$\left| \int k(x_0, y) - k(x_1, y) dg(y) \right| \leq \epsilon .$$

It is obvious that if one removes a finite number of rational points f where $(f, Tg_{n_k}) \to (f, Tg_0)$ does not hold, then the remaining rational points are still weakly separable. The only point which would cause difficulty would be $x = 0$ but $k(0, y)$ is continuous in y so that (f, Tg_{n_k}) converges to (f, Tg_0). It is to be observed that since T and $\bar{T}$ in this example satisfy symmetrical conditions, we obtain that

$$\min_g \max_f (f, Tg) = \max_f \min_g (f, Tg) .$$

EXAMPLE 2. Let

$$k(x, y) = \begin{cases} M(x, y) & x < y \\ M'(x, y) & x \geq y \end{cases} .$$

The same conclusion holds where in the preceding argument the right is replaced by the left.

EXAMPLE 3. The curve of discontinuity can be replaced by $y = \varphi(x)$ which is continuous monotonic $\varphi(0) = 0$ $\varphi(1) = 1$.

EXAMPLE 4. Let $\varphi(x)$, $\psi(x)$ be strictly increasing, and $\psi(x) > \varphi(x)$ for $0 < x < 1$ $\varphi(0) = \psi(0) = 0$ $\varphi(1) = \psi(1) = 1$. Let

$$k(x, y) = \begin{cases} M(x, y) & 1 \geq y \geq \psi(x) \\ M'(x, y) & \psi(x) > y \geq \varphi(x) \\ M''(x, y) & \varphi(x) > y \geq 0 \end{cases} ,$$

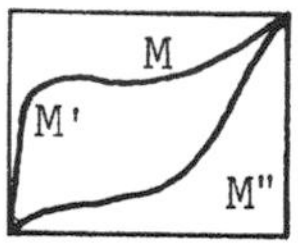

then the conclusion of Theorem 5 holds.

REMARK 2. In the proof of the existence of the value of game it is sufficient to use one sided conditions. Definitions 5 and 6 can be replaced by the conditions:

(a) For any sequence g_n there exists a subsequence g_{n_k} and g_0 such that for any $\epsilon > 0$ for every f, with the exception of at most a finite number, there exists an $M_0(f, \epsilon)$ such that for $n_k \geq M_0$, we obtain

$$\epsilon \geq (f, Tg_0) - (f, Tg_{n_k}) .$$

(b) There exists a countable sequence f_i such that for each g there exists an $f_i(g)$ depending on ϵ and g such that

$$\epsilon \geq (f, Tg) - (f_i, Tg)$$

with the analogous extension given in Definition 7.

REMARK 3. In terms of a specific example the above extension implies that if

$$K(x, y) = \begin{cases} M(x, y) & x < y \\ \varphi(x) & x = y \quad 0 \leq x, y \leq 1 \\ M'(x, y) & x > y \end{cases}$$

where $M(x, y)$, $\varphi(x)$ and $M'(x, y)$ are respectively continuous in the regions $x \leq y$, $x = y$, $x \geq y$, then the value of the game exists provided that

(*) $M(x, x) \geq \varphi(x) \geq M'(x, x)$ for $0 \leq x \leq 1$.

It is not known whether for such kernels the condition (*) can be strengthened. It seems that the only crucial points occur at the two end points (0, 0) and (1, 1).

The reason that the hypothesis is one sided and only imposed on T instead of on T and $\overline{T}$ is imbedded in the fact that if an operation T of a Banach space is w.c.c., then $T^* \supset \overline{T}$ (T^* the conjugate mapping to T) is automatically w.c.c.

We remark finally that many minor extensions to the min max theorem can be presented. In particular, the requirement of finite number in Definition 6 is not essential. It can be shown by simple examples that generally many of the results from Theorem 1 to Theorem 5 cannot be improved upon for the case of the unit square.

§ 2.

In this paragraph we proceed along the same direction as in Paragraph 1, and relate the method used to known results established by Wald and others. We summarize briefly the essential features of the first paragraph. Generally, two hypotheses were needed, a condition of weak complete continuity, and a requirement of weak separability. We dwell momentarily on the meaning of the concept "weak." In simple language it is not necessary to know beforehand an optimal strategy which will work uniformly against any defense of the opponent, but merely the knowledge of an effective response against any specific strategy used by the opponent. The effective response may vary and depend on the strategy chosen by player II. Thus, knowing an effective response for each strategy plus some sort of compactness (weak) guarantees that there will exist a uniform response. But for emphasis, we repeat that in the hypothesis only weak requirements are made.

We now turn to investigate the relationship of the two weak hypotheses.

DEFINITION 8. An operator T is said to be strongly separable if the range of Tg is separable.

DEFINITION 9a. An operator T is said to be strongly completely continuous (s.c.c.) if for any sequence g_n there exists a subsequence g_{n_k} and g_0 with

$$\lim (f, Tg_{n_k}) = (f, Tg_0)$$

uniformly for all f.

In the case where T is s.c.c. it is unnecessary to postulate any separability hypothesis.

THEOREM 6.[3] If T is s.c.c. then the space Tg may be topologized so that it becomes bicompact.

PROOF. A metric can be introduced into Tg as follows:

$$\rho(Tg_1, Tg_2) = ||Tg_1 - Tg_2|| = \sup_f (f, Tg_1 - Tg_2) .$$

The hypothesis on T implies that the space of Tg is sequentially compact and metric, and hence bicompact.

Thus in this case an appeal directly to Theorem 3 yields the max min theorem. However, if we do not suppose the existence of the limit point g_0 in Definition 9a, then an appeal to Theorem 3 cannot be made.

DEFINITION 9b. An operator is said to be completely continuous (c.c.) if for any sequence g_n there exists a subsequence g_{n_k} where $\lim (f, Tg_{n_k})$ exists uniformly in f. The g_0 of Definition 9a need not exist.

REMARK 4. It T satisfies the hypothesis of Definition 9b, and in addition the space of f was weakly separable (see Definition 5), then a reexamination of the proof of Lemma 5 will show that on account of the uniformity of the convergence of (f, Tg_{n_k}), the lemma still remains valid. Hence Theorem 1 also holds, provided that min sup is replaced by inf sup.

---

[3]The author is indebted to the referee for this observation.

It is interesting to observe that if T satisfies the conditions of Definition 9b, then weak separability is automatically achieved, therefore, the remark above applies. We now demonstrate the weak separability.

THEOREM 7. If T is c.c., then there exists a countable sequence of f_n such that for every g and f_0 there exists a f_n with

$$|(f_n, Tg) - (f_0, Tg)| \leq \epsilon .$$

PROOF. Since the image of Tg is compact in the uniform topology (every infinite sequence has a uniformly convergent subsequence), the range of Tg is separable uniformly with respect to f. Indeed, since the range of Tg is compact it is totally bounded and hence for every $\frac{1}{n}$ there exists a finite number of $Tg_{\alpha(1)}, Tg_{\alpha(2)}, \ldots, Tg_{\alpha(n)}$ such that for any g there exists a $Tg_{\alpha(i)}$ with

$$|(f, Tg) - (f, Tg_{\alpha(i)})| \leq \frac{1}{n} \quad \text{for all } f.$$

Collecting all the $Tg_{\alpha(i)}$ for every n establishes the separability of the range of Tg. Let Tg_n denote the dense set. If we consider the set of points (f, Tg_i) for fixed i then there exists a sequence of f_n^i which are dense in the values of (f, Tg_i). Let

$$\{f_k\} = \{f_m^i\}_{i,m} .$$

It is easy to verify now that f_k is dense in the sense described in the theorem. To this end, let f and g be given, then there exists a g_n with

$$|(f, Tg) - (f, Tg_n)| \leq \epsilon$$

for all f. Furthermore, we have

$$|(f, Tg_n) - (f_m, Tg_n)| \leq \epsilon .$$

Combining, we secure

$$|(f, Tg) - (f_m^n, Tg)| \leq 3\epsilon . \qquad \text{Q.E.D.}$$

We now verify that if $K(x, y)$ is continuous for $0 \leq x, y \leq 1$, then the transformation

$$h(x) = Tg = \int K(x, y) dg(y)$$

is completely continuous. Let g_n denote a sequence of distribution functions. Invoking Helly's selection theorem, we can choose a subsequence of $g_{n_k} = g_r$ with $g_r(y) \longrightarrow g_0(y)$ at every point of continuity of $g_0(y)$.

We demonstrate that

$$||Tg_r - Tg_s|| = \max_{0 \leq x \leq 1} |h_r(x) - h_s(x)| \longrightarrow 0 .$$

If we suppose the contrary, then there exists a sequence x_v with $|h_{r_v}(x_v) - h_{s_v}(x_v)| \geq \epsilon_0$ for a fixed ϵ_0. We select a subsequence of x_v convergent to a point x_0. For convenience, we suppose $x_v \longrightarrow x_0$. Clearly,

$$|h_{r_v}(x_0)-h_{s_v}(x_0)| \geq |h_{r_v}(x_v)-h_{s_v}(x_v)| - |h_{r_v}(x_v)-h_{r_v}(x_0)| - |h_{s_v}(x_v)-h_{s_v}(x_0)|.$$

But in view of the continuity of $K(x, y)$

$$|h(x_v) - h(x_0)| \leq \int |K(x_v, y) - K(x_0, y)| \, dg(y) \leq \frac{\epsilon_0}{4}$$

for v chosen sufficiently large independent of g.

Thus for $v \geq v_0$

$$|h_{r_v}(x_0) - h_{s_v}(x_0)| \geq \epsilon_0 - \frac{\epsilon_0}{4} - \frac{\epsilon_0}{4} = \frac{\epsilon_0}{2} .$$

But since $g_r \longrightarrow g_0$ and $K(x, y)$ is continuous, we conclude that $h_r(x) \longrightarrow h_0(x)$ for every x. This contradiction yields the assertion. Thus, we have shown:

> THEOREM 8. If $K(x, y)$ is a continuous function defined on the unit square, then the transformation
>
> $$Tf = \int K(x, y) \, df(x)$$
>
> is completely continuous mapping the distribution functions defined on the unit interval into the space of continuous functions defined on the unit interval.

It was observed by Wald [4], that one can achieve a min max theorem without restricting oneself to the special underlying space of the unit square. We turn now to investigate the relationship of Wald's results and the methods employed here. Let X and Y denote two arbitrary sets of points. Let $K(x, y)$ denote a bounded function defined on the product space $X \otimes Y$. A metric is introduced in X as follows:

$$\rho(x_1, x_2) = \sup_y |K(x_1, y) - K(x_2, y)| .$$

It is understood that we have introduced equivalence classes by identifying the points x_1 and x_2 where $\rho(x_1, x_2) = 0$. It is not necessary to assume that X is complete subject to the metric defined above. The space Y is handled similarly. Wald's criterion states that for a given ϵ there exists a finite number of x_i such that for every x there corresponds an x_i with

$$|K(x, y) - K(x_1, y)| \leq \epsilon$$

uniformly in y. This assumption implies that X is conditionally sequentially compact (every infinite sequence has a convergent subsequence). In what follows we assume that the Wald criterion on X is fulfilled. Since X is metric and compact, it follows that X is also separable. Let $C(X)$ and $C(Y)$ denote the Banach space of all continuous bounded functions defined on X and Y respectively. Let E_1 and E_2 denote their conjugate spaces. It is well known [5] that every element of E_1 corresponds to a finitely additive measure. Namely, if f is in E_1 and φ is in $C(X)$, we have

$$f(\varphi) = \int_X \varphi(x)\, m_f(dx) .$$

The natural partial ordering in $C(X)$ generates a partial ordering in E_1 such that the positive elements of E_1 correspond to the positive measures of E_1. An element f will be called a distribution if $f \geq 0$ and $||f|| = 1$. This means that f is associated with a positive measure and has variation equal to 1. The points of x correspond to the pure distributions.

Since X is separable and compact metric, it is easy to verify that $C(X)$ is a separable Banach space. We now demonstrate that E_1 is weakly compact as functionals. This shall play the role of the Helly's Theorem in this general case. In fact, generally the conjugate space of any separable Banach space B is always weakly compact as functionals. To this end, let $y_1, y_2, y_3, \ldots$ be a dense set in B. If f_k are a sequence of functionals uniformly bounded $(||f_k|| \leq c)$, then by the diagonal process we select a subsequence of f_k which converge for every y_m. For convenience, we let f_k denote the subsequence. If y is an arbitrary element of B, then let y_{m_0} be chosen so that $||y - y_{m_0}|| \leq \epsilon$. Thus

$$|f_i(y) - f_k(y)| \leq |f_i(y - y_{m_0})| + |f_k(y - y_{m_0})| + |(f_i - f_k)(y_{m_0})|$$

$$\leq 2\, c\, ||y - y_{m_0}|| + |f_i(y_{m_0}) - f_k(y_{m_0})| .$$

Hence, $\overline{\lim}\, |f_i(y) - f_k(y)| \leq 2\, \epsilon\, c$. This shows that f_k converges weakly as functionals. Let f_0 denote the limit functional.

In what follows, it will not be necessary to establish the weak convergence of a sequence of distributions over all of $C(Y)$, but only over a separable subspace of Y. We must recall the fact that the Wald criterion was not imposed on Y, and hence $C(Y)$ is not, a priori, necessarily separable. However, we restrict ourselves to a separable subspace L of $C(Y)$ consisting of all the functions $\varphi_x(y) = K(x, y)$. The Wald criterion placed upon X implies that the set of $\varphi_x(y)$ in $C(Y)$ is separable. The weak convergence with respect to L of any set of distributions can be secured as in the discussion above. It is, however, clear that no limit element as a functional is defined since L is not a closed linear subspace of C(Y). In the circumstance here, since the transformation $\int K(x, y)m_g(dy) = Tg$ is a completely continuous transformation (to be shown below), it is unnecessary to postulate the existence of a weak limit element in the set of distributions (see Remark 4).

Finally, we show that Tg is completely continuous. From the definition of the metric on X and Y, it is clear that K(x, y) is in C(Y) for each fixed x. Now let g_j be a collection of distributions. We may assume that g_j is weakly convergent over the space of functions $\varphi_x(y) = K(x, y)$.

The Wald criterion guarantees us for any given ϵ the existence of a finite number $x_1, \ldots, x_n$ which are total bounded with respect to all x. We select j and k sufficiently large so that

$$|\int K(x_i, y)[m_{g_j}(dy) - m_{g_k}(dy)]| \leq \epsilon$$

for $i = 1, \ldots, n$. This is possible by virtue of the weak convergence of g_k. Since for any x there exists an x_i such that

$$\sup_y |K(x, y) - K(x_i, y)| \leq \epsilon ,$$

we obtain

$$||Tg_j - Tg_k|| = \sup_x |\int K(x, y)[m_{g_j}(dy) - m_{g_k}(dy)]| \leq 3\epsilon$$

or

$$\max_x |Tg_j(x) - Tg_k(x)| \leq 3\epsilon .$$

It is observed in the proof that it is only necessary to impose the Wald criterion on one space X to achieve the result. Thus, summing up:

THEOREM 9. If K(x, y) satisfies the Wald criterion, then the transformation

$$Tg = \int K(x, y)\, m_g(dy)$$

defined over all distributions on Y with Y metrised as

$$\rho(y_1, y_2) = \sup_x |K(x, y_1) - K(x, y_2)|$$

is a completely continuous mapping of E into C(X).

It should be emphasized that the distributions obtained here are not necessarily representable in the usual manner as non-decreasing functions.

We close this section with two observations that, first, since Tg is a completely continuous operator, T^*f is also completely continuous. In particular the operator

$$\overline{T}f = \int K(x, y) m_f(dx)$$

which is contained in T^*, is also completely continuous. Thus the image of the pure distributions is a compact set in the space of bounded functions, or that the Wald Criterion holds automatically in Y if it was merely assumed to hold only in X. Thus the reason that only one sided conditions are needed, lies in the relations that the operator T has with T^*.

Now that we know that both spaces are compact, then an alternative procedure to the above is to complete the space X to $\overline{X}$. The space Y is treated similarly. They remain compact metric. Now, $C(X)$ shall consist of all uniformly continuous functions defined on X. In view of the compactness of both X and Y, $K(x, y) = \varphi_y(x)$ is in $C(X)$, the same applies to $C(Y)$.

Clearly $C(X) \simeq C(\overline{X})$ that is, the Banach space of bounded uniformly continuous functions over X is indistinguishable from that over $\overline{X}$. Clearly, the properties of total boundedness in x uniform in y of $K(x, y)$ are maintained. Since $C(X) \simeq C(\overline{X})$, we obtain that the conjugate space E_1 over both are identical. The importance here is that the representation of elements of E_1 correspond to completely additive measures defined over the Borel field of sets [6]. The variation of any positive measure is obtained by operating on the element 1 of $C(\overline{X})$ i.e., $f(1) = ||f||$. From this it follows easily that a weakly convergent sequence of distributions converges to a distribution. Now since both $C(X)$ and $C(Y)$ are separable, E_1 and E_2 are both weakly sequentially complete compact spaces as functionals.

Since the conjugate space over $C(X)$ and $C(\overline{X})$ is the same, the inf sup theorem can be replaced by a max min theorem. The essential need was the weak completeness as functionals of the conjugate space; whereas before we could only conclude the weak convergence of a set of distributions for a certain set of elements in $C(Y)$. The fact that we achieve max min is an essential strengthening of the result of Wald.

Furthermore, in this case the game need not be played with respect to finitely additive measures, but over all countably additive measures and the value will exist. A more detailed analysis of this phenomenon will be given in the next section.

General Considerations. In this last section we shall discuss generally the essential features of the previous investigation dealing with the Wald criterion. Let X and Y be given as before on each of which is given a Borel collection of sets $\mathfrak{A}$ and $\mathfrak{B}$ respectively. Let $L_1(X)$ and $L_2(Y)$ denote respectively the class of all measurable bounded functions over

X and Y. (A function $\varphi(x)$ is said to be measurable over X if for any real number α the set $A = [x | \varphi(x) < \alpha]$ is in $\mathfrak{U}$.) In what follows, it will be assumed that $\varphi(x) \equiv 1$ and $K(x, y) = \varphi_y(x)$ are in $L_1(X)$ for every y. This is a very essential requirement. The same holds for $L_2(Y)$. A few examples of $\mathfrak{U}$ over X are as follows:

(1) Let $\rho(x_1, x_2) = \sup_y |k(x_1, y) - k(x_2, y)|$ and $\mathfrak{U}$ shall consist of the smallest countably additive class of sets containing the open sets generated by the metric and the full space.

(2) Let a neighborhood of x_0 in X be defined as follows:

$$V(x_0; y_1, \ldots, y, \epsilon) = [x| \; |k(x, y_i) - k(x_0, y_i)| < \epsilon] \quad i = 1, \ldots, n .$$

Let $\mathfrak{U}$ again be the smallest countably additive class of sets containing these open sets and the full space.

(3) Let a countably additive measure m be defined on a Borel field of sets $\mathfrak{F}$ of X with respect to which 1 and $k(x, y) = \varphi_y(x)$ are measurable. Let $\mathfrak{U}$ be the set of all measurable sets with respect to the measure m.

Now, $L_1(X)$ and $L_2(Y)$ can be topologized in many different ways. Each topology on $L_1(X)$ generates a class of linear functionals continuous with respect to the topology on $L_1(X)$. Let E_1 and E_2 denote the functional spaces over $L_1(X)$ and $L_2(Y)$ respectively. The non-negative elements of E_1 and E_2 which have the value 1 for the element $\varphi(x) \equiv \psi(y) \equiv 1$ will be referred to as distributions which will be denoted as F and G respectively. Again, we present several examples. All our discussion will be confined to $L_1(X)$, but it is to be understood that $L_2(Y)$ is treated similarly.

(a) If $L_1(X)$ is topologized by the uniform norm, namely;

$$||\varphi|| = \sup_x |\varphi(x)| ,$$

then E_1 consists of all finitely additive set functions of bounded variation defined over $\mathfrak{U}$.

The next two examples are formulated in terms of convergence instead of topologies. Weak neighborhood topologies could have also been used.

(b) A sequence of elements $\varphi_n(x)$ in $L_1(x)$ converges to $\varphi(x)$ if for every x $\varphi_n(x) \longrightarrow \varphi(x)$ and $|\varphi_n(x)| \leq k$. This defines a convergence in $L_1(X)$ of which E_1 consists of all countably additive set functions of bounded variation defined over $\mathfrak{U}$.

(c) If $\mathfrak{U}$ is as in example 3, then $\varphi_n(x)$ converges to $\varphi(x)$ if $\varphi_n(x) \longrightarrow \varphi(x)$ almost everywhere and $|\varphi_n(x)| \leq k$. This convergence in $L_1(X)$ generates for E_1 the set of all summable functions with respect to the measure m.

It is observed that each time the functional space is different. In each circumstance a different game can be introduced with the kernel defined as follows:

$$(*) \qquad M(F, G) = F_x[G_y(k(x, y))] = G_y[F_x(k(x, y))] ,$$

where F and G are distributions respectively in E_1 and E_2. In all the examples given below $M(F, G)$ is well defined and the game can thus be played. The truth of (*) depends upon a Fubini theorem or thus the requirement of joint measurable of $k(x, y)$ in x and y. It is interesting to notice that in examples (a) and (b) the pure strategies are elements of E_1, while in (c) the points of X are not in E_1. Furthermore, in (a) and (b) the pure strategies are the extreme points for the set of distribution and in a weak sense convexly span all distributions. This is essentially the celebrated Krein Milman extreme point theorem [7], or it follows directly from the definition of integrals involved.

This easily gives that in these cases

$$\overline{v} = \inf_G \sup_F M(F, G) = \inf_G \sup_x M(x, G)$$

$$\underline{v} = \sup_F \inf_G M(F, G) = \sup_F \inf_y M(F, y) ,$$

where x and y denote typical pure distributions. Under this relationship, we obtain that if E_1 under topology 1 contains E_1 generated by topology 2, this implies that $\overline{v}^{1} \leq \overline{v}^{2}$. A similar remark holds for E_2, of course, with the inequality reversed for $\underline{v}$. This states that if under a second topology you generated more functionals than under the first topology and the value of the game existed under the first topology, then it must exist under the second topology. In other words, the more functionals you can generate, the more likely that the game will have a value. It will now be shown for the case of the unit square that every bounded measurable function possesses a value with respect to the game generated by the uniform topology. The game is played over all finitely additive distributions in place of countably additive distributions.

More precisely:

THEOREM 10. If $k(x, y)$ is bounded measurable in both variables and measurable for each x as a function in y, and for each y as a function in x. Let $L_1(X)$ and $L_2(Y)$ be given as above and topologized under the uniform norm $(||\varphi|| = \sup_x |\varphi(x)|)$ and if E_1 and E_2 denote the set of all finitely additive set function over X and Y respectively. (Let F and G denote distributions in E_1 and E_2), then

$$\max_F \min_G \int_0^1 \int_0^1 k(x, y)\, F(dx)\, G(dy) = \min_G \max_F \int_0^1 \int_0^1 k(x, y)\, F(dx)\, G(dy)$$

REMARK 5. This result depends upon the well known theorem that the unit sphere of the conjugate space of a Banach space in the weak topology as functionals (weak * topology) is bicompact [8]. In particular, the set of distributions being a closed sub-space of the unit sphere in the weak * topology is therefore also bicompact. This remark does not hold generally if the original space is not a Banach space.

PROOF. The proof of the theorem is now a simple consequence of this remark. Indeed, let F range over the pure distributions of E_1, thus

$$\iint k(x, y)\, F(dx)\, G(dy) = \int k(x_0, y)\, G(dy) = G(k(x_0, y))$$

$$= (\overline{T}F, G) = (F, TG) .$$

The remark above gives the bicompactness of the set of G in the weak * topology with respect to the distributions consisting of a finite number of jumps, and thus this is the bicompactness of the space of Remark 1.

Finally, we get by applying a simple adaptation of Theorem 3 that

$$\sup_F \min_G (F, TG) = \min_G \sup_F (F, TG) .$$

Since the hypotheses are symmetrical, max min can actually be achieved. Q.E.D.

The result of Theorem 10 can be secured for an arbitrary set X and Y by the natural extension.

In essence Theorem 10 has shown that every bounded measurable kernel generates a game with a value provided one plays the game over a suitable large functional space. The above procedure amounts to forming a weak * closure of the operator in the second conjugate space.

There is no inconsistancy as far as the continuous kernels are concerned. If one restricts oneself to the Banach space of all continuous functions over the unit interval instead of the larger space of bounded measurable functions $L_1(X)$, then the conjugate space is isomorphic to the set of all functions of bounded variation continuous to the right (V) (countably additive measures). One secures again a value for the game generated by the continuous kernel played over the countably additive measures (functions of bounded variation continuous to the right). The reason being that Remark 5 applies, in this case, as V is the conjugate space of $C(0, 1)$ which contains $k(x, y) = \varphi_x(y)$ where k is continuous.

Thus, one can restrict oneself to the smallest Banach space $B \subset L_1(X)$ containing $\varphi_x(y) = k(x, y)$ (the same holds for the Y space) and play the game over the distributions of the conjugate space to this Banach space. However, if one would like to play a game with respect to the countably additive measures E_1 over a kernel $k(x, y)$ which is not continuous, then no longer are the elements of E_1 acting as functionals over the space of functions $\varphi_x(y) = k(x, y)$, but as elements upon which the functionals $\varphi_x(y)$ are operating. The validity of Remark 5 does not hold for weak * topology replaced by weak topology, and thus to insure the value of the game, stronger hypotheses are needed of $k(x, y)$ to obtain the needed compactness. As for some examples, we refer to weak complete continuity of the operator generated by $k(x, y)$.

As to example (c) it is of a different type. The points x are no longer admitted as distributions. This is connected with the fact that the integrable functions do not constitute a conjugate space of a Banach space. Nevertheless, an analysis can be made to obtain criteria for the existence of a value of the game:

$$\iint k(x, y)\, \varphi(x)\, \psi(y)\, dx\, dy$$

where $\varphi(x) \geq 0$, $\psi(y) \geq 0$ almost everywhere and $\int \varphi(x)\, dx = \int \psi(y)\, dy = 1$. This suggests that an abstract max min can be developed for an arbitrary Banach lattice. This will be published elsewhere.

BIBLIOGRAPHY

[1] MORGENSTERN, O., von NEUMANN, J., "Theory of Games and Economic Behavior," 2nd ed. (1947), Princeton University Press, Princeton.

[2] VILLE, J., "Traité du Calcul des Probabilités et de ses Applications," Emil Borel and collaborators, Vol. 2, Part 5 (1938), Paris.

[3] VILLE, J., "Sur la Théorie Générale des Jeux où Intervient L'habilité des Joueurs," Traité du Calcul des Probabilités et de ses Applications, Emil Borel and collaborators, Vol. 2, Part 5 (1938), Paris, pp. 105-113.

[4] WALD, A., "Foundations of a General Theory of Sequential Decision Functions," Econometrica, Vol. 15 (1947), October.

[5] MARKOFF, A., "On Mean Values and Exterior Densities," Sbornik (1938).

[6] KAKUTANI, S., "Concrete Representations of (M) Spaces," Annals of Math. (1941).

[7] MILMAN, D., KREIN, M., "On Extreme Points of Regular Convex Sets," Studia Mathematica (1940).

[8] ALAOGLU, L., "Weak Topologies of Normed Linear Spaces," Annals of Math. (1940).

S. Karlin

The RAND Corporation

ON A THEOREM OF VILLE

H. F. Bohnenblust and S. Karlin[1]

§1. INTRODUCTION

Among the several procedures which lead to the existence of the value of a discrete two person, zero sum game — one is based on a theorem of Ville [1] and another is based on a fix point theorem of Kakutani [2]. In the present paper these two theorems are extended under certain conditions to infinite dimensional spaces. The results are useful tools in the theory of non-discrete games.

The theorem of Ville deals with matrices a_{ik}, $i = 1, \ldots, m$; $k = 1, \ldots, n$. It states that if to each $q_k \geq 0$, $\sum q_k = 1$ there exists corresponding $p_i \geq 0$, $\sum p_i = 1$ such that the bilinear form $\sum a_{ik}p_i q_k$ is non-negative, then there exists fixed $p_i \geq 0$, $\sum p_i = 1$ such that for all $q_k \geq 0$, $\sum q_k = 1$ the bilinear form is non-negative. In the generalization the $\{p_i\}$ are considered as elements of a Banach space B_1, and $\{q_k\}$ as elements of a Banach space B_2. In both spaces a cone of positive elements is assumed given. The matrix is replaced by a mapping of B_1 into the conjugate space B_2^* of B_2. The value of the bilinear form (Ap, q) is defined and it is investigated under what conditions the formal statement of Ville's theorem can be carried over to this abstract situation.

The generalization of Kakutani's theorem extends this theorem along the same lines as Schauder's fixed point theorem [3] generalizes the classical theorem of Brouwer.

§2. REGULARLY CONVEX SETS

Let B be a Banach space and B^* its conjugate space. The elements of B are denoted by $x, y, \ldots$ and those of B^* by $f, g, \ldots$. The value of the functional f at x is written in the form (f, x). The w^* - topology in B^* is the weak topology induced by the elements of B as functionals. A set F of B^* is called regularly convex if to each $f_0 \notin F$ there exists an element x_0 such that

$$(f_0, x_0) > \sup(f, x_0) \quad \text{over} \quad f \in F .$$

[1]Received June 8, 1949 by the ANNALS OF MATHEMATICS and accepted for publication; transferred by mutual consent to ANNALS OF MATHEMATICS STUDY No. 24.

A regularly convex set is convex and w^* - closed. A partial converse theorem states that convexity and w^* - compactness of a set implies regular convexity. Boundedness and w^* - closure imply w^* - compactness. Finally the following theorem is recalled:

> THEOREM 1. If F and G are regularly convex, disjoint sets of B^* and if one of them is w^* - compact then there exists an element x_0 and a real number α such that
>
> $$(f, x_0) > \alpha \text{ for all } f \text{ in } F;$$
>
> $$(g, x_0) < \alpha \text{ for all } g \text{ in } G.$$
>
> In other words, there exists a plane $(f, x_0) = \alpha$ which separates the sets F and G.

A set X in B is called a cone if $x \in X$ implies $\lambda x \in X$ for every real, non-negative number λ. A convex, closed cone X determines a cone F in B^* by:

$$F = \{f \mid (f, x) \geq 0 \text{ for all } x \text{ in } X\} . \tag{1}$$

The cone F is evidently regularly convex. Conversely, the cone F determines the closed convex cone X in the sense that $x \in X$ if and only if $(f, x) \geq 0$ for all f in F. If $x \in X$ this is true by the definition of F. On the other hand if $x_0 \notin X$, a functional f_0 and a real number α exist such that $(f_0, x_0) < \alpha$ and $(f_0, x) > \alpha$ for all x in X, since X is assumed closed and convex. But since $0 \in X$, α must be negative. Furthermore, $(f_0, x) \geq 0$ for every $x \in X$, since otherwise $(f_0, \lambda x)$ would be $< \alpha$ for λ sufficiently large. Hence f_0 must belong to F, and for this f_0, the value (f_0, x_0) is negative.

> THEOREM 2. Let G be a w^* - compact regularly convex set in B^* and X a closed convex cone in B with the property:
>
> to each $x \in X$ there exists a g of G for which $(g, x) \geq 0$.
>
> Then there exists a $g_0 \in G$ such that for all $x \in X$, $(g_0, x) \geq 0$.

PROOF. Let F be the cone in B^* determined by X according to (1). Theorem 2 states simply that F and G must intersect. Assume by contradiction that F and G have no point in common. Then by Theorem 1, there exists x_0 and a real number α such that

$$(f, x_0) > \alpha \text{ for all } f \text{ in } F, \ (g, x_0) < \alpha \text{ for all } g \text{ in } G .$$

Since F contains 0, the number α must be negative. For no f in F can (f, x_0) be negative, for otherwise for some large λ, $(\lambda f, x_0)$ would be $< \alpha$. Thus x_0 must belong to X, and since $(g, x_0) < \alpha < 0$ the assumption of Theorem 2 is contradicted. The assumptions on G in the preceding theorem can be weakened. Before stating the stronger theorem the following definition is introduced:

DEFINITION. A set H of B^* is said to satisfy property I if

(a) the set H is w^* - compact,

(b) the cone $\lambda H, \lambda \geq 0$ is convex,

(c) no segment joining two points of H contains the origin.

LEMMA 1. If a set H satisfies conditions (a) and (b) then the cone λH is regularly convex.

PROOF. If $f_0 \notin H$ and h is any element of H an element x exists such that $(f_0, x) < 0$ and $(h, x) > 0$. Since H is assumed w^* - compact, there exists a finite number of elements $x_1, \ldots, x_n$, such that $(f_0, x_i) < 0$ for each i and (h_i, x_i) are not all negative for each $h \in H$. The mapping of B^* into the n-dimensional space

$$f \longrightarrow \{\xi_1, \ldots, \xi_n\} \quad \text{where} \quad \xi_i = (f, x_i)$$

carries the cone H into a convex cone which does not intersect the "octant" $\xi_i < 0$, whereas the image of f_0 lies precisely in that octant. There exists $\mu_i \geq 0$, $\Sigma u_i = 1$ such that the plane $\Sigma \mu_i \xi_i = 0$ separates the image of f_0 from the image of the cone λH. For the element $x = \Sigma \mu_i x_i$ the inequalities $(f_0, x) < 0$ and $(f, x) \geq 0$ hold for each f in the cone λH. In other words, f_0 and λH are separated according to the definition of regular convexity.

LEMMA 2. If H satisfies property I there exists an element x_0 and a $\delta > 0$ such that $(h, x_0) \geq \delta > 0$ for all $h \in H$.

PROOF. By Lemma 1 the cone λH is regularly convex. Condition (c) of property I shows that $-h$ is not in the cone λH for $h \in H$. To each h_0 of H there exists an x and an α such that

$$(h_0, x) > -\alpha \quad \text{and} \quad (\lambda h, x) > \alpha$$

for each $\lambda \geq 0$ and $h \in H$. The real number α must be negative and hence (h_0, x) positive. Furthermore $(h, x) \geq 0$ for all h of H.

Again by compactness there exists $x_1, \ldots, x_n$ such that for each h of H, every $(h, x_i) \geq 0$ and at least one is positive. The element $x_0 = x_1 + \ldots + x_n$ satisfies the requirements in Lemma 2.

LEMMA 3. If H satisfies property I there exists a w^* - compact, regularly convex set G such that to any $h \in H$ there exists a $\lambda > 0$ for which $\lambda h \in G$ and conversely to any $g \in G$, a $\lambda > 0$ for which $\lambda g \in H$.

PROOF. Determine x_0 and $\delta > 0$ by Lemma 2. Let G be the intersection of the cone λH with the plane $(f, x_0) = 1$. The set G is regularly convex as the intersection of two regularly convex sets. If $h \in H$ then the element $h/(h, x_0)$ is in G. Conversely, if $g \in G$ then $g \neq 0$ and is in the cone λH. Thus for some λ, $\lambda^{-1} g$ is in H. Finally, G is w^* - compact since open sets covering G can be projected into open sets covering H.

THEOREM 3. Let H satisfy property I and let X be a closed convex cone in B with the property:

to each $x \in X$ there exists an $h \in H$ for which $(h, x) \geq 0$.

There exists then a fixed $h_0 \in H$ such that for all $x \in X$, $(h_0, x) \geq 0$.

PROOF. By Lemma 3 the set H can be replaced by a w^* - compact regularly convex set G to which Theorem 2 is applied.

§3. THE GENERALIZATION OF THEOREM OF VILLE

This theorem, stated in the introduction, is a simple corollary to Theorem 2 or Theorem 3. The image G of the set P of $\{p_i\}$, $p_i \geq 0$, $\sum p_i = 1$ under the transformation A is a convex, closed, bounded set in the n-dimensional space. The finite dimensionality of this space implies that G is regularly convex and Theorem 2 can be applied. The generalization of this theorem is natural. Let B_1 be a Banach space in which a set P is distinguished. Let A be a mapping of P into the conjugate space B_2^* of a Banach space B_2. In this second space a set Q is also distinguished. It is assumed that to each $q \in Q$ there exists a $p \in P$ such that $(Tp, q) \geq 0$. Theorem 3 asserts the existence of a $p_0 \in P$ for which $(Tp_0, q) \geq 0$ for all $q \in Q$ if

(1) the image of P under T satisfies property I,

(2) the cone λQ is convex and closed.

The second condition is satisfied for example, if Q is convex and weakly compact.

Many of the transformations arising in the theory of games satisfy conditions (1) and (2). The following case will serve as an illustration. Let B_1 and B_2 be the space of functions of bounded variation on the unit interval. Let P and Q, in B_1 and B_2 respectively, be the set of distribution functions, i.e., the set of non-decreasing functions of total variation 1.

Let $k(x, y)$ be a continuous function of x and y in $0 \leq x, y \leq 1$. The transformation of B_1 into B_2^* given by the integral $\int k(x, y)\, d\, G(x)$ is easily shown to be of the desired type.

§4. THE GENERALIZATION OF THE FIXED POINT THEOREM OF KAKUTANI

This theorem was proved for finite dimensional spaces. It states that if S is a convex, compact region in E_n and if to each $x \in S$ there corresponds a non-void set $A(x)$, convex and contained in S and such that $x_n \rightarrow x$, $y_n \rightarrow y$, $y_n \in A(x_n)$ imply $y \in A(x)$, then there exists a point $x_0 \in S$ which is contained in $A(x_0)$. This theorem is generalized to general Banach spaces under suitable compactness conditions.

THEOREM 4. Let S be a convex closed set of a Banach space. To each point x of S a non-void set $A(x) \subset S$ is assumed given. If

(a) $x_n \rightarrow x$; $y_n \rightarrow y$; $y_n \in A(x_n)$ imply $y \in A(x)$,

(b) if the union $\bigcup A(x)$ over all $x \in S$ is contained in some sequentially compact set T,

then there exists a point $x_0 \in S$ such that $x_0 \in A(x_0)$.

PROOF. The intersection of S and T is non-void and sequentially compact. Given $\epsilon > 0$ determine $s_1, \ldots, s_n$ in it such that these points are ϵ-dense in $S \cap T$. Let S_0 be the convex determined by the s_i. By construction the distance from the set S_0 of any point $y \in A(x)$ is less than ϵ. Denote by $A(x)$ the set of all points in the space which which are at a distance less than or equal to ϵ from $A(x)$. The sets $A(x)$ are closed and convex and the preceding remark shows that the closed, convex intersection of S_0 and $A(x)$ is non-void. Denote it by $B(x)$, it lies in S_0. If $x_n \rightarrow x$ and $z_n \rightarrow z$, $z_n \in B(x_n)$ then by construction there exist $y_n \in A(x_n)$ such that $||y_n - z_n|| \leq \epsilon$. But the y_n are in the sequentially compact set T and thus we may assume that $y_n \rightarrow y$.

Obviously $||y - z|| \leq \epsilon$ and $y \in A(x)$. Thus $z \in B(x)$ and Kakutani's theorem can be applied to the set S_0 with the correspondence $x \rightarrow B(x)$. This leads to the result, that to each $\epsilon > 0$ there exists an $x \in S$ (in fact in S_0) and a $y \in A(x)$ such that $||x - y|| \leq \epsilon$.

Pick a sequence $\epsilon_n \rightarrow 0$, determine x_n, y_n in the above manner. The y_n may be assumed to converge, say, to y. Since $||x_n - y_n|| \leq \epsilon_n$ the x_n converge also to $y \in A(y)$ by the hypothesis (a).

THEOREM 5. Let B be a Banach space whose conjugate space B^* is weakly separable. Let S be a convex, w - closed set of B. Assume that to each $x \in S$ there corresponds a non-void set $A(x) \subset S$ such that

(a) $A(x)$ is convex,

(b) $x_n \rightarrow_w x$, $y_n \rightarrow_w y$, $y_n \in A(x_n)$ imply $y \in A(x)$,

(c) the union $\cup A(x)$ is contained in some sequentially w - compact set T,

then there exists an $x_0 \in S$ such that $x \in A(x_0)$.

PROOF. Let E be space of bounded sequences $\{\xi_n\}$ with the norm: $\sup |\xi_n|$. Let f_n be a weakly dense set in B^*. The mapping U: $\xi_n = (f_n, x)/n$ of B into E carries convex sets into convex sets, and if $x_n \rightarrow_w x$ then $Ux \rightarrow Ux$. Furthermore $Ux_1 = Ux_2$ implies $x_1 = x_2$. Applying the mapping U to S and to each $A(x)$ leads us in the space E · and all the conditions of Theorem 4 are satisfied. Theorem 5 becomes a simple consequence of Theorem 4.

BIBLIOGRAPHY

[1] VILLE, J., "Sur la Théorie Générale des Jeux où Intervient L'habilité des Joueurs," Borel Collection, Paris, Vol. 2, No. 5 (1938), pp. 105-113.

[2] KAKUTANI, S., "A Generalization of Brouwer's Fixed Point Theorem," Duke Mathematical Journal (1941), pp. 457-459.

[3] SCHAUDER, J., "Der Fixpunktsatz in Funktionalräumen," Studia Mathematica, Vol. 2 (1930), pp. 171-180.

H. F. Bohnenblust

S. Karlin

The RAND Corporation

POLYNOMIAL GAMES[1]

M. Dresher, S. Karlin, L. S. Shapley[2]

§0. INTRODUCTION

A basis is laid in this paper for a theory of two-person zero-sum games in which the payoff is a polynomial function $P(x, y)$ of the two strategy variables x and y, the latter taking their values from closed, one-dimensional intervals. A somewhat more general category of "polynomial-like" games is examined first: games whose payoff has the form

$$K(x, y) = \sum_{i=1}^{m} \sum_{j=1}^{n} a_{ij} r_i(x) s_j(y) ,$$

r_i and s_j being any continuous functions. A general discussion of games with continua of strategies appears elsewhere in this volume [2].

Polynomial games are important as a bridge, leading from the discrete games, whose theory has been well explored, to more general classes of infinite games which admit polynomial approximations to their payoff functions. No nontrivial properties of such approximations have been obtained. One immediate observation is that the error in the game value does not exceed the least upper bound of the error in the payoff. A similar uniformity in the approach to the optimal strategies can not be guaranteed in general. The approximative properties of polynomial-like games are presumably superior in some respects to those of polynomial games; but, as will be seen, the results achieved here for the wider class are less sharp.

The feature of polynomial and polynomial-like games which links them to the discrete case is the finite dimensionality of the spaces of mixed strategies. We may study the solutions of a discrete, $m \times n$ game by means of a geometric model involving an $m - 1$-dimensional simplex and an n-dimensional convex polyhedral cone (the positive orthant of Euclidean n-space). The corresponding figures for a polynomial game, of degrees m and n in x and y respectively, are the m-dimensional moment space and the $n + 1$-dimensional cone of the nth degree polynomials non-negative over the interval $0 \leq y \leq 1$.

---

[1]Portions of this paper were presented to the American Mathematical Society at Columbus, Ohio in December 1948, and at Palo Alto, California, in April 1949.

[2]Received September 13, 1949 by the ANNALS OF MATHEMATICS and accepted for publication; transferred by mutual consent to ANNALS OF MATHEMATICS STUDY No. 24.

An extended study of the geometrical properties of the moment spaces and of the non-negative polynomial spaces is to be published elsewhere [6]; a review of many of the results has already appeared [5]. In the present paper the relevant portions are cited (with informal proofs) (§4) and applied (§5) to derive a number of inequalities relating the sets of optimal mixed strategies of the two players in a polynomial game. The results (Theorems 6 and 7) are not as sharp as corresponding results in the discrete case [1], [4] because the moment spaces and the polynomial cones are not polyhedral figures. It seems likely that some new indices characterising these spaces, beyond those considered in this paper, will have to be introduced before the inequalities can be substantially improved.

A possible computational procedure for polynomial games is outlined in the final section (§6).

§1. CONICAL RECIPROCATION IN GAME THEORY

A finite dimensional, bilinear game may be described as follows: Player I chooses a point $r = (r_1, \ldots, r_m)$ from a set R lying in Euclidean m-space. Player II chooses a point $s = (s_1, \ldots, s_n)$ from a set S in Euclidean n-space. R and S are bounded, closed, convex. The kernel $A(r, s)$, or payoff from II to I, is given by the matrix (a_{ij}) in the form

$$A(r, s) = \sum_{i,j=1}^{m,n} a_{ij} r_i s_j . \tag{1}$$

The minimax theorem for bilinear functions over convex sets [7] asserts that

$$\min_{s \in S} \max_{r \in R} A(r, s) = \max_{r \in R} \min_{s \in S} A(r, s) = v , \tag{2}$$

thereby defining the <u>value</u> v of the game. <u>Optimal strategies</u> for the two players are defined to be points r^o, s^o such that

$$\min_{s \in S} A(r^o, s) = v, \quad \max_{r \in R} A(r, s^o) = v . \tag{3}$$

The sets of optimal strategies, as functions of the coefficients (a_{ij}), vary in a semi-continuous manner, described in the following theorem.

THEOREM 1. Let G be an open set containing the set $R^o(A)$ of optimal strategies of the first player in the game A. Then $\varepsilon = \varepsilon(G) > 0$ exists such that $R^o(B)$ is contained in G for every game B with coefficients (b_{ij}) satisfying

$$|b_{ij} - a_{ij}| \leq \varepsilon \quad \text{all } i, j .$$

(Compare Lemma 6 in [1].)

PROOF. Suppose the contrary. Then a sequence $\{B^{(k)}\}$ of games can be found with

$$|b_{ij}^{(k)} - a_{ij}| \leq \varepsilon_k \quad \{\varepsilon_k\} \longrightarrow 0$$

each of which has an optimal strategy $r^{(k)}$ outside of G. The value $v^{(k)}$ of $B^{(k)}$ converges to a value of A. The $r^{(k)}$ have a limit point in the compact region $R - G$. Since each $r^{(k)}$ guarantees the amount $v^{(k)}$ in the game $B^{(k)}$ this limit point is seen to be an optimal strategy in the original game A, and hence lies in $R^{O}(A) \subset G$. This contradiction proves the theorem.

We now describe a principle of conical reciprocation which gives us a geometrical approach to the study of the structure of the solutions. This method was used previously in the analysis of discrete and convex kernels [1], [2], [4]. Imbed the set S in $n + 1$-space by affixing the coordinate $s_0 = 1$ to each point. Construct from it the closed, convex cone P_S of points λs, $s \in S$, $\lambda \geq 0$. The reciprocal, or conjugate cone, P_S^*, is defined to be the set of points $h = (h_0, \ldots, h_n)$ for which

$$\sum_{j=0}^{n} h_j s_j \geq 0 \quad \text{all} \quad s \quad \text{in} \quad P_S \; .$$

P_S^* is a convex, closed cone, and the fundamental duality theorem [9] states that $(P_S^*)^* = P_S \cdot P_R$ and P_R^* will denote the analogous cones in $m + 1$-space generated by the set R.

The way to visualize the connection between P_S and P_S^* is to consider the latter as a region in the space of all oriented hyperplanes through the origin of $n + 1$-space. These hyperplanes may be represented by the homogeneous linear functions

$$H(s) = \sum_{j=0}^{n} h_j s_j = 0, \quad \text{some} \quad h_j \neq 0$$

positive multiples of H being identified. Using the orientation, we may regard them as half-spaces. Then P_S^* is essentially the set of closed half-spaces that contain P_S. The interior of P_S^* is the set of open half-spaces containing P_S (except for the origin). The boundary points of P_S^* are just the supporting hyperplanes to P_S. The points $(h_0, \ldots, h_n)$ themselves lie on the directed normals to the hyperplanes they determine. This description may of course be dualized to give P_S in terms of P_S^*.

We now define:

$$
(4) \qquad \begin{aligned} a_{00} &= -v, \\ a_{01} = \cdots &= a_{0n} = 0, \\ a_{10} = \cdots &= a_{m0} = 0 . \end{aligned}
$$

The effect of thus augmenting the matrix (a_{ij}) is merely to change the value of the game from v to 0, and to adapt the form (1) to the higher dimensional spaces in which the cones are constructed.

For convenience we introduce the operator A and the inner product notation (h, s) for points in $n + 1$-space, giving us

$$
(Ar, s) \quad \text{for} \quad \sum_{j=0}^{n} \Big(\sum_{i=0}^{m} a_{ij} r_i \Big) s_j .
$$

Let AR stand for the image of R under A, plotted in $n + 1$-space. Since AR lies within the hyperplane $h_0 = -v$ its dimension is not more than n. (See however § 4.)

The dimension of a convex set of Euclidean n-space is the dimension of the smallest linear manifold containing the set. An interior point is one having a (full n-dimensional) neighborhood entirely within the set; the other points are the boundary points. By an inner point of a p-dimensional convex set we shall mean one that has a neighborhood whose intersection with the set is open in the p-dimensional manifold in which the convex set is contained: an inner point is interior if and only if $p = n$. An inner point can be represented as a convex combination of points in the set in such a way that any preassigned point of the set occurs with positive weight.

A hyperplane separates two convex sets if the two closed half-spaces determined by the hyperplane each contain one of the sets. Two convex sets can always be separated if they have no inner point in common (but this is not a necessary condition for separation).

LEMMA 1. The convex bodies AR and P_S^* intersect; in fact

$$
AR \cap P_S^* = AR^0 ,
$$

if R^0 denotes the set of optimal strategies of player I.

PROOF. By (3) and (4) any r^0 in R^0 satisfies

$$
(Ar^0, s) \geq 0 \quad \text{all } s \text{ in } S.
$$

It follows that Ar^0 is a point of P_S^*. Hence

$$
AR^0 \subset AR \cap P_S^* .
$$

Conversely, for any r in R, if Ar is in P_S^* then

$$(Ar, s) \geq 0 \quad \text{all } s \text{ in } S ,$$

and hence such an r is optimal for player I. Thus

$$AR \cap P_S^* \subset AR^0 .$$

The proof is completed with the observation that the existence of optimal strategies assures that AR^0 is not empty.

LEMMA 2. The convex bodies AR and P_S^* can be separated by a hyperplane; in fact, the separating hyperplanes correspond one-one with the optimal strategies of player II.

PROOF. Let S^0 denote the set of optimal strategies of player II. Every s^0 in S^0 satisfies

$$(Ar, s^0) \leq 0 \quad \text{all } r \text{ in } R,$$

$$(h, s^0) \geq 0 \quad \text{all } h \text{ in } P_S^* .$$

Thus s^0 represents a separating hyperplane. Conversely, consider any separating hyperplane H. Since the two convex bodies are themselves in contact (Lemma 1), H is a plane of support to both. A plane of support to a cone necessarily contains the vertex of the cone: in the case of P_S^* this point is the origin. Hence we may represent H as a homogeneous linear functional $s = (s_0, \ldots, s_n)$, some $s_j \neq 0$, such that $(h, s) = 0$ for each point h of H. With proper choice of sign we may then write:

$$(5) \qquad (Ar, s) \leq 0 \quad \text{all } Ar \text{ in } AR,$$

$$(6) \qquad (h, s) \geq 0 \quad \text{all } h \text{ in } P_S^* .$$

By (6), s is a point of $(P_S^*)^* = P_S$. Since $s \neq 0$, normalization by a positive factor will yield a unique point in the bounded cross section S of the cone P_S. By (5) s is a point of P_{S^0}, so that the normalized point is an optimal strategy of player II. Clearly, distinct planes lead to distinct strategies, and vice versa, making the correspondence biunique.

The two lemmas (and their obvious counterparts in terms of SA and P_R^*) give a geometric significance to the sets of optimal strategies, and will enable us to establish a variety of dimensional relationships between them. For sharp results we shall require detailed information about the boundaries of the several convex bodies involved.

§2. GEOMETRIC MODEL OF THE DISCRETE GAME

If R is taken as the simplex $r_i \geq 0$, $i = 0, 1, \ldots, m$, in the m-dimensional hyperplane

$$\sum_{i=0}^{m} r_i = 1$$

of $m + 1$-space, then the cone P_R is the positive orthant $r_i \geq 0$ of $m + 1$-space. The cone is self-reciprocal: $P_R^* = P_R$. If S, n, etc., are taken similarly, then we obtain a geometric model of the general zero-valued game with finite sets of strategies. Because the regions here are polyhedral an exact accounting of the dimensions of the optimal strategy sets can be given [1], [4].

§ 3. POLYNOMIAL-LIKE GAMES

We shall be concerned henceforth with games defined by a kernel of the following form:

$$K(x, y) = \sum_{i,j=1}^{m,n} a_{ij} \, r_i(x) s_j(y) \tag{7}$$

where the functions r_i and s_j are continuous, and x and y, the strategies of players I and II respectively, are to be chosen from the intervals $0 \leq x \leq 1$, $0 \leq y \leq 1$.

Obviously, any payoff of the form

$$K(x, y) = \sum_{\nu=1}^{N} A_\nu(x) B_\nu(y)$$

can be put in the form (7), and conversely. The form (7) permits the greatest variety in a geometric model with dimensionality restricted by fixed m and n. The terminology "polynomial-like," "moments," is intended to suggest that the r_i and s_j be regarded as sets of orthogonal polynomials, or trigonometric functions, or simply as the powers x^i, y^j. (The last-named specialization is of course the ultimate object of the present paper.) However no such limitation on the natures of $r_i(x)$ and $s_j(y)$ will actually be demanded.

Mixed strategies may be represented by cumulative probability distribution functions $f(x)$ and $g(y)$, for players I and II respectively, characterized by being monotonic increasing and continuous to the right, with $f(-0) = g(-0) = 0$, $f(1) = g(1) = 1$. The mixed strategy payoff is then written as the Stieltjes integral:

$$(8)\quad \int_0^1 \int_0^1 K(x, y)df(x)dg(y) = \sum_{i,j=1}^{m,n} a_{ij} \int_0^1 r_i(x)df(x) \int_0^1 s_j(y)dg(y) .$$

Clearly, the only significant properties of the distribution functions are their "moments:"

$$(9)\qquad r_i = \int_0^1 r_i(x)df(x) \quad s_j = \int_0^1 s_j(y)dg(y)$$

for $i = 1, \ldots, m$, $j = 1, \ldots, n$. The substitution of (9) into (8) reveals that the polynomial game is equivalent to the bilinear game (1) of §1, provided that the sets R and S are convex.

We now characterize the "moment spaces" R and S. The theorem is stated only for R, the set S being entirely analogous. We shall need the following lemma, established by Fenchel [3].

LEMMA 3. If D is the convex closure of an arbitrary set C in n-space, then every point of D may be represented as a convex combination of at most n + 1 points of C. If C is connected, then not more than n points are necessary.

THEOREM 2. R is the convex set spanned by the curve C traced out in m dimensions by

$$r_i = r_i(x)$$

as x varies between 0 and 1.

PROOF. Let D be the convex set spanned by C, and suppose a point r^o of R, is not in D. Then there exists some hyperplane h strictly separating r^o from D. That is, for some fixed $\delta > 0$,

$$(10)\qquad \sum_{i=1}^{m} h_i r_i^o - \sum_{i=1}^{m} h_i\, r_i(x) \geq \delta$$

for any x in $0 \leq x \leq 1$. Since r^o is in R there is a distribution function $f^o(x)$ having the "moments" r_i^o. Integrating both sides of (10) with respect to $f^o(x)$ we obtain

$$(11)\qquad \sum_{i=1}^{m} h_i r_i^o \int_0^1 df^o(x) - \sum_{i=1}^{m} h_i \int_0^1 r_i(x)df^o(x) \geq \delta \int_0^1 df^o(x) ,$$

the inequality holding good because (10) is true for all x in the range of the integration. But with the aid of (9), (11) reduces to

$$\sum_{i=1}^{m} h_i r_i^o - \sum_{i=1}^{m} h_i r_i^o \geq \delta ,$$

giving the contradiction $0 > 0$. This proves that all points of R are in D.

Conversely suppose r^0 to be in D, and thereby to have a convex representation

$$r_i^0 = \sum_{k=1}^{N} \alpha_k^0 r_i(x_k) \quad i = 1, 2, \ldots, m,$$

where the α_k^0 are positive and add up to 1, with $N \leq m$ by virtue of Lemma 3. It is easily seen that the distribution

$$f^0(x) = \sum_{k=1}^{N} \alpha_k^0 I(x - x_k)$$

has the moments r_i^0, where $I(x - x_k)$ denotes the distribution function putting full weight on $x = x_k$. Hence every point D is in R, and the theorem is established.

THEOREM 3. In the polynomial-like game described, both players have optimal mixed strategies with at most min (m, n) steps.

PROOF (on player I). In the preceding proof it was observed that every point of R corresponds to a step-function distribution with $N \leq m$ steps. On the other hand, since AR is convexly spanned in n dimensions by the connected set AC, each point of AR is the image under A of a point of R spanned by at most n points of C. By Lemma 1 any r in R whose image is a point of $AR \cap P_S^*$ is optimal, hence some mixed strategy having not more than min (m, n) steps is optimal.

COROLLARY. If, in place of (7),

$$K(x, y) = \sum_{j=1}^{\infty} \sum_{i=1}^{m} a_{ij}\, r_i(x) s_j(y),$$

the convergence being uniform in y, then both players possess optimal mixed strategies with at most m jumps.

PROOF. As a result of the uniform convergence, the functions

$$s_i'(y) = \sum_{j=1}^{\infty} a_{ij} s_j(y) \quad i = 1, 2, \ldots, m$$

are continuous. The kernel may therefore be rewritten:

$$K(x, y) = \sum_{i=1}^{m} r_i(x) s_i'(y) = \sum_{i,j=1}^{m} \delta_{ij} r_i(x) s_j'(y),$$

(δ_{ij}) being the unit matrix, and the conclusion of the corollary follows from the theorem.

Letting both sums become infinite destroys in general the discrete nature of the optimal mixed strategies. L. J. Savage has pointed out a class of analytic kernels for which the unique optimal mixed strategies are absolutely continuous distribution functions.

THEOREM 4. If the dimensions (in R and S) of the sets of optimal mixed strategies are μ and ν respectively, and if ρ is the rank of the matrix (a_{ij}), $(i = 1, \ldots, m,\ j = 1, \ldots, n)$; then there exists an optimal mixed strategy for player I with at most

$$\min(\rho,\ n - \nu + 1)$$

steps, and for player II with at most

$$\min(\rho,\ m - \mu + 1)$$

steps.

PROOF (on player I): As in the proof of the preceding theorem we analyze the convex representation in AR of the points of contact with the cone P_S^*. The convex set AR is ρ-dimensional; hence by Lemma 3 every point is spanned by at most ρ points of the connected set AC. Furthermore every point of the contact set $AR \cap P_S^*$ lies in the $n - \nu$-dimensional intersection L of the hyperplanes separating the two bodies. The contact set must be contained in the convex closure of $AC \cap L$. Applying the weaker form of Lemma 3, since $AC \cap L$ may not be connected, we obtain a convex representation of any contact point by at most $\min(\rho,\ n - \nu + 1)$ points of AC. The rest of the proof is now evident. This theorem includes Theorem 3, since $\rho \leq \min(m, n)$.

THEOREM 5. In general

$$\mu + \nu \leq m + n - \rho,$$

where μ, ν, and ρ are defined as in Theorem 4.

PROOF. The set R^o of optimal points in R has dimension μ. The dimension of AR^o is at least $\mu - (m - \rho)$, the original dimension less the maximum possible loss due to the degeneracy of the transformation A. (That is, A is capable of collapsing an $m - \rho$-dimensional set into a point but nothing more.) On the other hand, AR^o is the contact set $AR \cap P_S^*$, and lies in $\nu + 1$ linearly independent hyperplanes in $n + 1$-space, whose intersection has dimension $n - \nu$. Hence

$$\mu - (m - \rho) \leq \dim (AR^{o}) \leq n - \nu .$$

COROLLARY. If the matrix (a_{ij}), $i = 1, \ldots, m$, $j = 1, \ldots, n$, is not degenerate, then

$$\mu + \nu \leq \max (m, n) .$$

§ 4. POLYNOMIAL GAMES: DESCRIPTION OF THE MOMENT SPACES

The polynomial game with kernel

$$K(x, y) = \sum_{i,j=0}^{m,n} a_{ij} x^{i} y^{j} \tag{12}$$

played on the unit square $0 \leq x \leq 1$, $0 \leq y \leq 1$ is a specialization of (7) of particular interest. Since $r_0 = s_0 = 1$ identically for all distribution functions, R and S are m- and n-dimensional regions already naturally embedded in $m + 1$- and $n + 1$-space. After some preparatory discussion we shall give a description in detail of these regions and of their conjugate cones P_R^* and P_S^*, paying special attention to the boundaries as they affect the possible contact sets.

The appearance of coefficients a_{i0} and a_{0j} not zero has not significantly altered the model. The statement of § 2 that AR has dimension at most n is no longer valid: the full number $n + 1$ of dimensions is now possible provided that $n < m$. As before we assume an adjustment of a_{00} to make the value of the game zero.

We define two indices of surface dimension, $a(x)$ and $c(x)$, for a point x in the boundary of a convex set D in n-space, in order to describe the nature of the hypersurface at that point. Let $L(x)$ denote the intersection of all the hyperplanes of support to D that contain x. Let $a(x)$ denote the dimension of $L(x)$, and let $c(x)$ denote the dimension of the convex set in which $L(x)$ meets D. The sets of points for which $a(x) = a$, $0 \leq a < n$, are the a-dimensional components, or "faces," of the boundary of D. The c-index tells in how many directions the boundary is actually flat. (Thus if D is polyhedral then $a(x) = c(x)$ everywhere.) Both indices are affine-invariant, and are not changed by increasing the dimension of the space in which the convex set is imbedded. This last fact suggests the definition $a(x) = c(x) = n$ for points x interior to D.

We shall have occasion to use $a(x)$ and $c(x)$ referring to points x in the boundary of a convex cone. In such context we calculate with respect to a bounded cross section through the cone at x, rather than with respect to the cone itself. The values of course are independent of the choice of cross section. For the vertex of the cone $a(x) = c(x) = -1$.

If D^* is a bounded cross section of the cone P_D^*, then the points y in the boundary of D^* may be regarded as the supporting planes to D, and vice versa. If $y \in D^*$ is an inner point of all y that are supporting planes to D at a point x, then we say y is <u>conjugate</u> to x. The relationship is not always symmetric: x will be a supporting plane to D^* at y, but not necessarily an inner supporting plane. In the figure, y is conjugate to x, but x is not conjugate to anything.

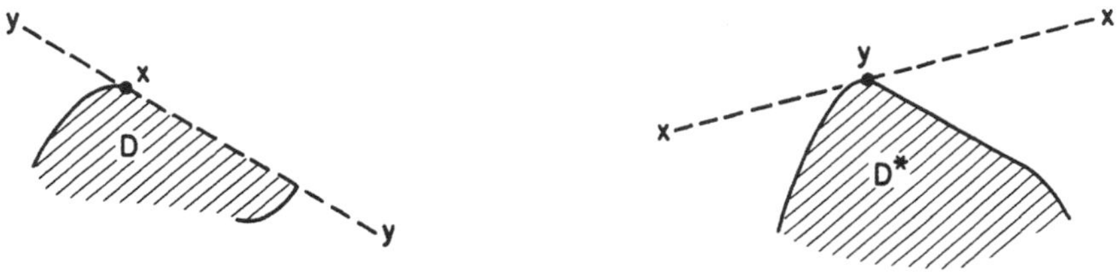

LEMMA 4. Let x be in the boundary of a convex set D in n-space, and let y be in the boundary of D^*, a bounded cross section of P_D^*. If y is conjugate to x, or x conjugate to y, then,

(13) $$a(x) + c(y) = c(x) + a(y) = n - 1 .$$

PROOF. Take any $x \in D$, $y \in D^*$ with $(x, y) = 0$. Let A be the set of planes of support to D^* at y:

$$A = \{x' \in D | (x', y) = 0\} .$$

There are $n - a(y)$ linearly independent x' in A. The dimension of A, considered as a convex subset in the boundary of D, is therefore $n - a(y) - 1$. On the other hand, the set $L(x) \cap D = B$ has dimension $c(x)$ by definition; while every x' in B satisfies $(x', y) = 0$. Therefore A contains B. Now the hypothesis tells us that either

(a) y is an inner plane of support to D at x, or

(b) x is an inner point of A.

In case (a), y supports D in precisely the set B; that is, (x', y) is actually positive for any x' in $D - B$. Hence B contains A. In case (b), every supporting plane through x must contain all of A (otherwise A would have points on both sides of the plane). Hence $L(x)$ contains A and B contains A. It follows that $B = A$ and $\dim B = \dim A$. This gives the right hand equality of (13). The other follows by symmetry, since D is a bounded cross section of $(P_{D^*})^*$.

The following description of the finite dimensional moment spaces is drawn from [6]. The set R is the convex closure of the curve C traced out by

$$r_i = x^i \quad i = 1, 2, \ldots, m$$

as x varies between zero and one (Theorem 2). If $m \geq 2$ then all the points of C are actually extreme (not expressible as convex combinations of other points of R). These points correspond to the pure strategies $I(x - x')$.

Points in the boundary of R have unique representations as convex linear combinations of extreme points. To see this fact, consider the characteristic property of any supporting hyperplane to R:

$$(14) \qquad h_0 + \sum_{i=1}^{m} h_i r_i \geq 0 \quad \text{all } r \text{ in } R .$$

This is equivalent to

$$(15) \qquad h_0 + \sum_{i=1}^{m} h_i x^i \geq 0 \quad \text{all } x \text{ in } 0 \leq x \leq 1 .$$

The equality holds for at least one value of x, but not for more than m values, as not all the coefficients of the polynomial (15) vanish at once. Thus, the curve C does not touch the hyperplane more than m times. The $k \leq m$ contacts are linearly independent since their coordinates from the Vandermonde matrix

$$\begin{pmatrix} x_1 & x_2 & \cdots & x_k \\ x_1^2 & x_2^2 & & x_k^2 \\ \cdot & & & \\ \cdot & & & \\ \cdot & & & \\ x_1^m & x_2^m & \cdots & x_k^m \end{pmatrix}$$

whose rank is k. Every supporting hyperplane therefore meets the boundary of R in a simplex, the points of which can be represented in precisely one way as convex combinations of the vertices. But every point in the boundary of R is touched by some supporting hyperplane. Q.E.D.

The smallest number of points of C by which a point r can be spanned we shall denote by $b(r)$. We let $b'(r)$ denote the same number, but with the end points $(0, \ldots, 0)$ and $(1, \ldots, 1)$ counted conventionally as half points. Thus $b'(r)$ can take on half-integral values, and always

$$(16) \qquad b(r) - 1 \leq b'(r) \leq b(r) .$$

The previous discussion of the contact simplex has effectively established that

(17) $$b(r) = c(r) + 1$$

for r on the boundary of R. If we observe that any root in $0 < x < 1$ of the polynomial (15) must be double in order to preserve the inequality, it is then easy to show that

(18) $$2b'(r) = a(r) + 1 .$$

For r interior to R we have the following improvement on Lemma 3:

(19) $$2b'(r) = m + 1 ;$$

which suggests taking $a(r) = m$ in the interior. (This, together with $c(r) = m$, is the value obtained from the definitions if R is imbedded in a space of higher dimension. Formula (17) is not valid under this extension.) On the boundary of R we have, by (16), (17), (18):

$$a(r) - 1 \leq 2c(r) \leq a(r) + 1 .$$

That is, c is approximately half of a.

If the notion of representation by convex linear combinations is extended to permit the use of infinite sets of extreme points, then the convex representations of any r^0 in R correspond one-one to the distributions $f^0(x)$ whose moments are the coordinates of r^0.

The cone P_R is naturally obtained by considering the moment

$$f_0 = \int_0^1 df(x)$$

as an $m + 1$st coordinate r_0 and allowing distributions with arbitrary non-negative total weight $f_0 \geq 0$. The probability distributions of R have of course all had $f_0 = 1$.

The conjugate cone P_R^*, because of the close connection between the linear form (14) and the polynomial (15), can be regarded as the set of all polynomials, of degree m or less, which are not negative in the interval $0 \leq x \leq 1$. The boundary of P_R^* comprises just those with at least one root in the interval. The extreme points (of a bounded cross section of the cone) are just those with all m roots in the interval. The latter fall into two components according to whether the root 1 occurs with even or odd multiplicity; this is reflected in the sign of the leading coefficient h_m. Every point of the cross section lies on some line segment connecting the two components of extreme points; that is, every point is spanned by two extreme points. A convenient bounded cross section, which we call R^*, is obtained by the normalization:

$$\sum_{i=0}^{m} h_i/(i + 1) = \int_0^1 \sum_{i=0}^{m} h_i x^i \, dx = 1 .$$

Much of the foregoing material has been given for its conceptual interest, without proof, since it makes no contribution to the rigor of the derivations of the next section. Detailed proofs, and many ramifications and applications of this material, will appear in [6].

§ 5. POLYNOMIAL GAMES: THE SETS OF OPTIMAL STRATEGIES

We now proceed to derive the main results. Let μ and ν denote as before, the dimensions of the sets of optimal moment strategies for players I and II respectively. Let ρ be the rank of the $m \times n + 1$ matrix (a_{ij}) omitting $i = 0$; σ the rank of the $m + 1 \times n$ matrix (a_{ij}) omitting $j = 0$. (Note that the change in a_{00} to make the game value zero does not affect these ranks.) The dimension of AR is then precisely ρ. Our attention will be chiefly directed to the "general" case $\rho = \min(m, n + 1)$, $\sigma = \min(m + 1, n)$.

In particular, let $\rho = m$, so that the transformation A is non-singular. All the dimensional indices are preserved under A: $a(r) = a(Ar)$, $c(r) = c(Ar)$. Take an optimal point r^0 such that Ar^0 is an inner point of the contact set $AR \cap P_S^*$. Take s^0 to be an inner point of the set $S^0 \subseteq S$ of optimal strategies of player II. Ar^0 and s^0 are "almost always" conjugate. However it can happen that neither is conjugate to the other. To cover such cases we shall need a point on the surface of the cone that is conjugate to s^0; we shall call this point s^*. Such a point always exists if s^0 is in the boundary of S; if s^0 is interior we define the vertex 0 of the cone to be conjugate to s^0, observing that Lemma 4 is valid with the convention $a(s^0) = c(s^0) = n$, $a(0) = c(0) = -1$.

Several dimensional inequalities may be derived:

(a) Directly from the definitions we have

$$\mu \leq c(r^0), \quad \nu \leq c(s^0) . \tag{20}$$

(b) Every plane of support to P_S^* at s^* contains the contact set $AR \cap P_S^*$. But the dimension of the contact set is just μ. Hence, allowing for contact along the conical elements of P_S^*, we have $c(s^*) + 1 \geq \mu$, or, by Lemma 4,

$$\mu \leq n - a(s^0) . \tag{21}$$

(c) There are $n + 1 - a(r^0)$ independent supporting hyperplanes to AR at the point Ar^0. The ν-dimensional family of separating planes all support AR at that point; hence,

(22) $$\nu \leq n - a(r^0) ,$$

since the number of linearly independent items is one more than the dimension number.

(d) On the other hand, there are at least $n - a(s^*)$ independent hyperplanes through the origin supporting the cone P_S^* at the point Ar^0, since, as remarked in (b) above, every supporting plane at s^* contains the contact set. To weed out the hyperplanes that do not separate, we must only require that they support AR as well. This imposes at most $a(r^0) - \mu$ further constraints, leaving at least $n - a(s^*) - a(r^0) + \mu$ independent separating planes. Hence,

$$\nu \geq n - a(s^*) - a(r^0) + \mu - 1 ;$$

or, using Lemma 4,

(23) $$\mu - \nu \leq a(r^0) - c(s^0) .$$

(e) The intersection of the separating planes has dimension $n - \nu$. Each separating plane contains s^* and Ar^0, in the boundaries of P_S^* and AR respectively; and hence contains sets with dimension $c(s^*) + 1$ and $c(r^0)$. The common part of these sets is at most μ-dimensional. Therefore

$$n - \nu \geq c(s^*) + 1 + c(r^0) - \mu .$$

Using Lemma 4 this becomes:

(24) $$\mu - \nu \geq c(r^0) - a(s^0) .$$

Despite the symmetry of (23) and (24), separate derivations were necessitated by our one-sided hypothesis $\rho = m$.

The inequalities (20) through (24) may be converted with the help of (17)-(19) into statements about the more palpable (from the standpoint of the game) quantities $b(r^0)$, $b'(r^0)$, etc. For we may define a pure strategy x as essential if there is an optimal mixed strategy in which x is played [4]. Then if r^0 is in the boundary of R player I has precisely $b(r^0)$ essential pure strategies, by the uniqueness of the convex representations of boundary points. On the other hand, if r^0 is interior to R then every pure strategy is essential. We sum up:

THEOREM 6. In the polynomial game with kernel

$$K(x, y) = \sum_{i,j=0}^{m,n} a_{ij} x^i y^j \quad m \leq n + 1$$

in which the $m \times n + 1$ matrix (a_{ij}), $i \neq 0$, has rank m, the mixed strategies are representable as points in the m- and n-dimensional moment spaces R and S. If r^0 is an inner point of the μ-dimensional convex set of optimal moment strategies for player I; if similarly s^0 and ν for player II, then

$$\mu \leq n + 1 \quad - 2b'(s^0)$$
$$\nu \leq n + 1 \quad - 2b'(r^0)$$
$$\mu - \nu \leq 2b'(r^0) - b(s^0)$$
$$\nu - \mu \leq 2b'(s^0) - b(r^0) \quad .$$

Also, if r^0 is in the boundary of R, then

$$\mu \leq b(r^0) - 1$$

and the number of essential pure strategies for I is exactly $b(r^0)$. Similarly, if s^0 is in the boundary of S, then

$$\nu \leq b(s^0) - 1$$

and the number of essential pure strategies for II is $b(s^0)$.

The approximate equality (16) of b and b' should be recalled in interpreting this theorem.

Interior optimal moment strategies. The case in which player II has an optimal s^0 interior to S merits special notice. Here the origin of $n + 1$-space is the sole contact point of the two convex bodies. From (21) we now get the precise result $\mu = 0$, since $a(s^0) = c(s^0) = n$. Moreover the argument of (d) now counts exactly the number of constraints on the family of separating planes, and yields

$$\nu = n - a(r^0)$$

(which may also be derived from a sharpening of the argument of (c)). We have established:

THEOREM 7. In the polynomial game of Theorem 5, if s^0 is interior to S, then player I's optimal moment strategy r^0 is unique, and

$$\nu = n + 1 - 2b'(r^0) \quad .$$

Still retaining the one-sided condition $\rho = m$, we now suppose that player I has an interior optimal moment strategy. In this case m must be less than $n + 1$ in order to expose the inner points of AR to contact. But substituting $a(r^o) = c(r^o) = m$ into (20)-(24) does not lead beyond the inequalities of Theorem 6. In other words, the assumption of an interior optimal strategy, for the player with the greater number of dimensions to play with, is quite strong (Theorem 7); while the same assumption for the other player is hardly restrictive at all.

In the square case, $m = n = \rho$, if s^o is interior and unique, then r^o is also interior and unique, and

$$b'(r^o) = b'(s^o) = (n + 1)/2 .$$

There is a simple construction for games of this type, which is merely a matter of finding a nonsingular, $n + 1 \times n + 1$ matrix A such that

$$Ar^o = s^oA = (v, 0, \ldots, 0) ,$$

where r^o and s^o are the desired interior moment strategies and $v \neq 0$ is the desired value.

Equalizing optimal strategies. If it happens that one player has an optimal strategy which makes the outcome independent of his opponent's action, then a notable simplification occurs in the computation of the solution (see the next section). For example, the pure strategies x^o and y^o are equalizing optimal strategies if the payoff can be written in the form

$$K(x, y) = P(x)Q(y)R(x, y) + k ,$$

with x^o a root of P, y^o a root of Q. The geometrical significance, if player I has an optimal equalizer, is that the origin lies in AR. For player II it means that some plane of support to P_S^* contains the whole set AR. Existence of an interior solution for one player requires that all optimal strategies of his opponent be equalizers. For an interior moment strategy may be realized as a convex combination of pure strategies in which given pure strategy appears with positive weight. Use of this pure strategy against any opposing optimal strategy can only give the value of the game as payoff. Hence, every opposing optimal strategy must be an equalizer. (Compare [1], Lemma 2.) The converse is not true; all of one player's optimal strategies may be equalizers without the other player having an interior moment strategy which is optimal. This contrasts with a property of finite games: that only the essential pure strategies yield the value of the game against every opposing optimal strategy. ([1], Theorem 1.)

The foregoing considerations are valid as well for polynomial-like games, with "inner solution" for "interior solution," since the full dimensionality of the moment space does not figure in the argument. No rank restrictions need be placed on (a_{ij}).

We note in conclusion that an equalizing strategy is not a fortiori an optimal strategy. This fact somewhat reduces the value of the "equalizer" concept, so far as computation is concerned (see below).

§6. POLYNOMIAL GAMES: COMPUTATION

This brief section is included to indicate the order of magnitude of the computational difficulties, when the solution of a polynomial game is tackled directly. There is no discussion of approximation methods.

It is possible to reduce the solution of the game problem to the solution of certain systems of algebraic equations — linear in some cases, non-linear in the remaining.

(a) If there exists an equalizing $g^0(y)$ such that

$$\int_0^1 \sum_{i,j=0}^{m,n} a_{ij} x^i y^j dg^0(y) = w \quad \text{for all } x , \tag{25}$$

then, since the moment $g_0^0 = 1$, we have

$$a_{00} + \sum_{j=1}^{n} a_{0j} g_j^0 = w \tag{26}$$

$$a_{i0} + \sum_{j=1}^{n} a_{ij} g_j^0 = 0 \quad i = 1, 2, \ldots, m . \tag{27}$$

(The g_j here are of course the s_j of earlier sections. The emphasis is now on the distribution functions rather than on the geometry.) The equations (27), solved for g_j^0, cannot give the moments of an equalizer unless the rank ρ of the matrix (a_{ij}), $i \neq 0$, is n or less. More generally, if $\rho > n - \nu$ then no equalizing g^0 exists, as a simple contradiction shows, and it is useless to proceed with this attack.

A solution g_j^0 of (27) is a set of moments if and only if the two quadratic forms

$$\sum_{k,l=0}^{n'} g_{k+l}^0 x_k x_l, \quad \sum_{k,l=0}^{n'-1} (g_{k+l+1}^0 - g_{k+l+2}^0) x_k x_l \quad (\text{if } n = 2n') ,$$

$$\sum_{k,l=0}^{n'} g_{k+l+1}^0 x_k x_l, \quad \sum_{k,l=0}^{n'} (g_{k+l}^0 - g_{k+l+1}^0) x_k x_l \quad (\text{if } n = 2n' + 1) ,$$

are non-negative definite (compare [8], p. 77). The equalizer $g^0(y)$ is then a solution of the game only if the constant w, given by (26), is in fact the value of the game. This can ordinarily be established only by the discovery of an optimal $f^0(x)$. In particular, if the solution of

$$a_{0j} + \sum_{i=1}^{m} a_{ij}f_i^0 = 0 \quad j = 1, 2, \ldots, n$$

is a set of moments as well, then w is automatically the value, and the equalizers $f^0(x)$, $g^0(y)$ solve both the given game and its negative.

The problem of reconstructing a distribution function from a given (finite) set of moments is essentially equivalent to that of obtaining the roots of a certain polynomial, closely related to the quadratic forms above, whose degree is approximately half the number of moments given [6].

(b) In general, the process (a) will not solve the game unless equalizers exist for both players. If player II, say, has no equalizer, then the polynomial (25) can reach its maximum in $0 \leq x \leq 1$ at not more than $m' = [m/2] + 1$ points x_k (or $m/2$ points in the "b'" system of counting). The number of essential strategies for I is limited thereby to m', and his optimal distribution has the form:

$$f^0(x) = \sum_{k=1}^{m'} \alpha_k I(x - x_k) .$$

Non-uniqueness may appear in the α_k, but not in the x_k.

The sets of moments will have the form

$$f_i^0 = \sum_{k=1}^{m'} \alpha_k x_k^i , \quad i = 1, 2, \ldots, m ,$$

$$g_j^0 = \sum_{l=1}^{n'} \beta_l y_l^j , \quad j = 1, 2, \ldots, n ,$$

where $n' = [n/2] + 1$. The polynomials

$$H^0(x) = \sum_{i,j=0}^{m,n} a_{ij}g_j^0 x^i , \quad K^0(y) = \sum_{i,j=0}^{m,n} a_{ij}f_i^0 y^j$$

must satisfy the equations

$$\left.\begin{aligned} H^0(x_k) &= w_1 \\ K^0(y_l) &= w_2 \end{aligned}\right\} \quad \ldots\ldots\ldots \quad (\text{all } k, l) ,$$

$$\left.\begin{aligned} \frac{d}{dx} H^0(x_k) &= 0 \\ \frac{d}{dy} K^0(y_l) &= 0 \end{aligned}\right\} \quad \ldots\ldots\ldots \quad (\text{all } k, l \text{ except ends}) .$$

The four possible arrangements of essential strategies at the points 0 and 1 must be tried separately in solving these equations, and the condition on the derivative omitted in those cases where x_k or y_l is fixed equal to 0 or 1. In each case there turn out to be $m + n + 2$ unknowns α_k, β_l, x_k, y_l, and w_h. Together with the normalizations

$$\sum_{k=1}^{m'} \alpha_k = 1 \ , \ \sum_{l=1}^{n'} \beta_l = 1 \ ,$$

there are also $m + n + 2$ equations, which are linear in the α_k, β_l, w_h. Every solution of this system must give $w_1 = w_2$, as a simple argument shows. At least one solution exists in which all of the other unknowns lie between 0 and 1 (thereby giving legitimate distribution functions $f^o(x)$, $g^o(y)$), and having the "saddle point" property:

$$H^o(x) \leq w_1 = w_2 \leq K^o(y) \tag{28}$$

for all $0 \leq x \leq 1$, $0 \leq y \leq 1$. All solutions of this type are solutions of the game. In general there will be many other solutions of the equations which do everything but satisfy (28). These will locate maxima and minima of the original kernel (12), solve the negative of the given game, and perform other more obscure tasks. It is not until (28) that we make use of the primary motivations: of player I to maximize the kernel, and of player II to minimize.

In the foregoing treatment we have assumed the worst, i.e., that each player had the greatest possible (finite) number of essential pure strategies. A more practical approach might be to start with small values of m' and n', and work up.

BIBLIOGRAPHY

[1] BOHNENBLUST, H. F., KARLIN, S., SHAPLEY, L. S., "The Solutions of Discrete, Two-person Games," this Study.

[2] BOHNENBLUST, H. F., KARLIN, S., SHAPLEY, L. S., "Games with Continuous, Convex Pay-off," this Study.

[3] FENCHEL, W., "Krümmung und Windung Geschlossener Raumkurven," Math. Annalen 101 (1929), pp. 238-252.

[4] GALE, D., SHERMAN, S., "Solutions of Finite Two-person Games," this Study.

[5] KARLIN, S., SHAPLEY, L. S., "Geometry of Reduced Moment Spaces," Proc. N. A. S., 35 (1949), pp. 673-677.

[6] KARLIN, S., SHAPLEY, L. S., "Geometry of Finite Dimensional Moment Spaces," (to be published).

[7] von NEUMANN, J., "Über ein Ökonomisches Gleichungssystem und eine Verallgemeinerung des Brouwerschen Fixpunktsatzes," Ergenbnisse eines Mathematischen Kolloquiums, 8 (1937), pp. 73-83.

[8] SHOHAT, J. A., TAMARKIN, J. D., "The Problem of Moments," American Mathematical Society (1943), New York.

[9] WEYL, H., "Elementare Theorie der Konvexen Polyeder," Comentarii Mathematici Helvetici, 7 (1935), pp. 290-306. (See also this Study.)

M. Dresher

S. Karlin

L. S. Shapley

GAMES WITH CONTINUOUS, CONVEX PAY-OFF

H. F. Bohnenblust, S. Karlin, L. S. Shapley[1]

§1. BACKGROUND

In the "normal form" of a two-person, zero-sum game, as the theory has been set forth by von Neumann [3], there are just two moves. They are the choices of strategy, made simultaneously by each player. One player is then required to pay to the other an amount (positive or negative) determined by the pay-off function, which is a function only of the strategy-choices. The theory is best known at present for games in which the number of strategies available to each player is finite. This article will explore a rather special class of games in which the strategies of one player form a compact and convex region B of finite-dimensional Euclidean space, while those of the other form an arbitrary set A.

In general, equality may or may not hold in

$$\sup_{x \in A} \inf_{y \in B} M(x, y) \leq \inf_{y \in B} \sup_{x \in A} M(x, y) . \tag{1}$$

Intuitively, there may be a gap between what the x-player's best "safe" strategy guarantees to him and what his opponent's best "safe" strategy prevents him from obtaining. When equality does not hold, the typical procedure of game theory is to replace the choosing of a strategy by the choosing of a probability distribution over the whole set of strategies. Thus, the player entrusts the task of playing the game to a machine which makes random decisions, and contents himself with controlling its probable behavior to maximize his probable gain. Such a probability distribution is called a mixed strategy, and its order is the number of points in the spectrum of the distribution. (That is, the order is infinite unless a finite set of strategies exists which is chosen with probability one; in that case the order is the number of strategies which are chosen with positive probability.) A pure strategy is a mixed strategy of order one.

The game on the unit square will illustrate the use of mixed strategies without the inconvenient notation that general sets A and B would entail. Let $A = B$ be the closed one-dimensional unit interval $[0, 1]$. Then, corresponding to (1) is the inequality

[1] Received April 21, 1949 by the ANNALS OF MATHEMATICS and accepted for publication; transferred by mutual consent to ANNALS OF MATHEMATICS STUDY No. 24.

$$\sup_{F \in \mathcal{D}} \inf_{y \in B} \int_{-0}^{1} M(x, y)\,dF(x) \leq \inf_{G \in \mathcal{D}} \sup_{x \in A} \int_{-0}^{1} M(x, y)\,dG(y) \,, \tag{2}$$

where $\mathcal{D}$ is the set of all (cumulative) probability distributions on $[0, 1]$. ($F \in \mathcal{D}$ if and only if (i) $x < x_1$ implies $F(x) \leq F(x_1)$, (ii) $x < 0$ implies $F(x) = 0$, (iii) $x \geq 1$ implies $F(x) = 1$, (iv) F is continuous to the right.) Under quite general conditions, equality holds in the expression exemplified by (2), while not necessarily holding in (1). When it does hold, the number thereby defined is termed the value of the game. A distribution which achieves that value is termed an optimal mixed strategy (o.m.st.) for the player in question. Any pair of o.m.st. is termed a solution of the game. A game may in some cases have a value without having a solution.

§2. SUMMARY AND DISCUSSION OF RESULTS

A function φ is said to be convex if and only if, for any λ_1 and λ_2 satisfying

$$0 \leq \lambda_1 = 1 - \lambda_2 \leq 1 \,, \tag{3}$$

the inequality

$$\lambda_1 \varphi(x_1) + \lambda_2 \varphi(x_2) \geq \varphi(\lambda_1 x_1 + \lambda_2 x_2) \tag{4}$$

holds whenever all three terms are defined. It is strictly convex if, in addition, $x_1 \neq x_2$ and $\lambda_1 \lambda_2 \neq 0$ always imply the strict inequality in (4). The present paper deals with games in which the pay-off $M(x, y)$ is, for every x in A, a continuous convex function of y. Continuity in y and compactness of B are enough to assure the existence of a value, as has been shown by Wald [6]. Convexity in y further assures the existence of an optimal pure strategy for the y-player, that is, an o.m.st. of order one. The central result of the present paper is that the x-player must have an o.m.st. of order at most $n + 1$, where n is the dimension of B. Moreover, if the y-player has a p-dimensional set of o.m.st. of order one, then the x-player has an o.m.st. of order at most $n - p + 1$.

Without convexity the solutions, even of games on the unit square, may be much more complicated. If M if a polynomial, Dresher has shown that o.m.st. of finite order exist for both players [2]. But Blackwell and Girshick have found a unit square game with continuous pay-off in which the only o.m.st. for each player makes use of every strategy [2].

It might be worth-while to illustrate the way in which the results for convex games can be applied to other games. A linear function is of course convex, and the expected pay-off of a game is always linear in the mixed strategies. It follows that, in any game, if B is a finite set with m elements, then the x-player has an o.m.st. of order m or less. A more general statement is that if B can be subdivided into m closed, convex, non-overlapping components B_i, of dimension n_i, such that the pay-off is convex over each component, then the y-player has an o.m.st. of order at most m and the x-player one of order at most $m + \sum n_i$. The verification of either statement is accomplished by constructing an equivalent, convex game with an enlarged set B' of strategies for the y-player.

Symmetrically corresponding assertions obviously hold, here and throughout the paper, with concavity in x replacing convexity in y.

Thus consideration of convexity (concavity) is a handy tool for uncovering the existence of simple solutions in potentially complicated games. The question of computing such solutions when they exist will be discussed in paragraph 5 of the present paper.

§3. THEOREM ON CONVEX FUNCTIONS

Let B be a compact, convex region in an (n - 1)-dimensional space whose elements are denoted by y. A function f is <u>linear</u> (non-homogeneous) if $f(\sum \lambda_i y_i) = \sum \lambda_i f(y_i)$ when $\sum \lambda_i = 1$. The function $f(y) \equiv 1$, denoted by $\mathbf{1}$, is linear. The linear functions form an n-dimensional linear space E. F will denote an element of the conjugate space E^*.

LEMMA 1.1. If $F(\mathbf{1}) = 1$ there exists y such that $F(f) = f(y)$ for all f in E.

PROOF. It suffices to show that the n equations $F(f_i) = f_i(y)$ have a solution in y for n linearly independent elements f_i of E. But one may take $f_1 = \mathbf{1}$ and get an identity for the first equation. The remaining n - 1 equations, still independent, have a solution.

LEMMA 1.2. The set of all f which are non-negative over B forms a closed convex cone $P \subset E$, with vertex at the origin, containing $\mathbf{1}$ in its interior. Moreover, the region over which $P(y) \geq 0$ is precisely B.

(The notation "P(y)" will mean "f(y) for all f in P"; "f(B)" will mean "f(y) for all y in B.")

PROOF. The first part is obvious. For the latter, B and any y not in B can be separated; i.e., some f in E will have $f(y) < c \leq f(B)$. Then $f - c$ is in P and is negative for y.

LEMMA 1.3. Let Q be a compact convex set of E which does not intersect P. There exists y in B and δ such that $Q(y) \leq -\delta < 0$.

PROOF. P and Q are separated by some F in E^*; that is, for some $\delta > 0$, $F(Q) + \delta \leq F(P)$. Since P is a cone with vertex at the origin and F(P) is bounded from below, F(P) must be non-negative. Since $\mathbf{1}$ is in the interior of P, making $F(\mathbf{1}) > 0$, F may be chosen so that $F(\mathbf{1}) = 1$. By Lemma 1.1 a y exists satisfying

$$Q(y) + \delta \leq 0 \leq P(y) ,$$

while, by Lemma 1.2, y must be in the set B.

LEMMA 1.4. If $p_1, \ldots, p_m$ are points in an $(n - 1)$-dimensional space, then any point in their convex is in a convex spanned by at most n of them.

PROOF. Take a simplex S_m in $(m - 1)$-dimensional space and a linear transformation mapping it on the given convex C, the vertices of S_m going into the points $\{p_i\}$. The inverse transformation maps each point p of C onto a plane L(p), of dimension at least $m - n$, which intersects S_m. When a plane meets a simplex but not its boundary, the intersection is a point. Hence there is a simplicial face of S_m which intersects L(p) in a point. Its dimension must be less than n, and its vertices obviously correspond to a subset of $\{p_i\}$ which spans p.

LEMMA 1.5. If $\sup_\alpha f_\alpha(B)$ is positive for a family of $\{f_\alpha\}$, then for suitable $\lambda_i \geq 0, \sum \lambda_i = 1$, and α_i, $i = 1, \ldots, n$, the function $f = \sum \lambda_i f_{\alpha_i}$ is in P; that is, $f(B) > 0$.

PROOF. The Heine-Borel covering theorem permits one to work with a finite sub-family of $\{f_\alpha\}$, since $f_\alpha(y) > 0$ defines an open set. (This is the only use made of strict positiveness. The hypothesis might alternatively read "If $\sup_\alpha f_\alpha(B) \geq 0$ for a finite family ...". In this form Lemma 1.5 is equivalent to Ville's lemma [5].) The convex Q spanned by the finite sub-family must intersect P, by Lemma 1.3. Since Q is bounded and P is not, some boundary point of Q lies in P. This point is on a polyhedral face of dimension at most $n - 1$. Lemma 1.4 now gives us the desired representation.

THEOREM 1. Let $\{\varphi_\alpha\}$ be a family of continuous convex functions defined over a compact, convex, $(n - 1)$-dimensional region B. Then $\sup_\alpha \varphi_\alpha(y)$ attains its minimum value c at some point of B; and, given any $\delta > 0$,

$$\sum_{i=1}^{n} \lambda_i \varphi_{\alpha_i}(B) \geq c - \delta ,$$

for any suitable choice of α_i and $\lambda_i \geq 0$, $\sum \lambda_i = 1$.

PROOF. Let

$$a > \inf_{y \in B} \sup_{\alpha} \varphi_\alpha = c .$$

The set of y in B with $\varphi_\alpha(y) \leq a$ is non-void, closed and convex, and decreases as a decreases. The intersection of all these sets is non-void, and any point in it satisfies the first part of the theorem. For the second part, let $\{f_\beta\}$ be the family of linear functions with $[\varphi_\alpha - f_\beta](B) \geq 0$ for some α. This family contains all planes of support to all φ_α; therefore $\sup_\beta f_\beta = \sup_\alpha \varphi_\alpha$. Apply Lemma 1.5 to the family

$$\mathcal{F}_\delta = \{f_\beta - (c - \delta)\mathbf{1}\} , \ (\delta > 0) .$$

Each β_i so obtained corresponds to an α_i with $f_{\beta_i} \geq \varphi_{\alpha_i}$. These α_i and the λ_i of the lemma provide the representation of the theorem.

COROLLARY 1.1. If the (convex) set Y of points for which $\sup_\alpha \varphi_\alpha(y) = c$ has dimension p, then the number of functions φ_{α_i} required is at most $n - p$.

PROOF. Take an $(n - 1 - p)$-dimensional cross section $B' \subset B$ perpendicular to Y and intersecting Y in an interior point y_0. Let d_1 be the distance from y_0 to the nearest boundary point of Y, and let d_2 be the diameter of B. Apply the theorem to B' and $\delta' = \delta d_1/d_2$. The λ_i and α_i so obtained, $i = 1, \ldots, n - p$, must work for the original B and δ.

The following will be obtained in a somewhat different form in paragraph 7 and is put here for the sake of completeness.

COROLLARY 1.2. If Y is in the boundary of B, then the number of functions required is at most $n - p - 1$.

§4. APPLICATION TO GAMES

Consider a bounded pay-off function $M(x, y)$ where the choice x [y] of the maximizing [minimizing] player is taken from the set A [B]. M is continuous and convex in y for each x, and B is a compact, convex region in $(n - 1)$-dimensional Euclidean space. Let $Y \subseteq B$ denote the set of points which minimize $\sup_x M(x, y)$ and let p denote the dimension of Y. (Y is non-void, closed and convex.) Let I_z denote the pure strategy by which the point z is chosen with probability one. A mixed strategy will be called ε-_effective_ if the value of the game is not more than ε better than the expected return guaranteed by the mixed strategy to its user. Thus, an o.m.st. is 0-effective.

THEOREM 2. The value of the game described is

$$c = \min_{y \in B} \sup_{x \in A} M(x, y) .$$

For any $\varepsilon > 0$, there is an ε-effective mixed strategy for the x-player of the form

$$F_0 = \sum_{i=1}^{n-p} \lambda_i I_{x_i} \quad \left(\sum_{i=1}^{n-p} \lambda_i = 1, \lambda_i \geq 0\right) ; \tag{5}$$

while all pure strategies I_{y_0} on some y_0 in Y are optimal for the y-player.

PROOF. The theorem is a direct consequence of Theorem 1 and Corollary 1.1.

COROLLARY 2.1. If in addition A is compact and M is continuous in x for each y, then some mixed strategy of the form (5) is optimal.

PROOF. The added conditions make $\{M(\alpha, y)\}$ a closed family, hence it is permissible to take $\varepsilon = 0$ in the theorem.

COROLLARY 2.2. If, moreover, M is strictly convex in y for each x, then the y-player's o.m.st. is unique.

PROOF. Using some fixed o.m.st. $\sum \lambda_i I_{x_i}$, define Y_ν as the set of y with

$$\sum \lambda_i M(x_i, y) < c + 1/\nu \quad \nu = 1, 2, \ldots .$$

Let P denote the set function associated with any optimal y-strategy. Then it is easily seen that $P(B - Y_\nu) = 0$ for any ν. But strict convexity implies that $\bigcap Y_\nu$ is a single point y_0; hence the o.m.st. I_{y_0} is unique.

§ 5. COMPUTATION OF THE SOLUTION

Suppose, to avoid the complication of ϵ-effective mixed strategies, that the conditions of Corollary 2.1 are met, so that the game has an o.m.st. of the form (5). The determination of the value of the game

$$c = \min_{y \in B} \max_{x \in A} M(x, y)$$

and of the sets

$$Y = \text{those } y \text{ for which } \max_{x \in A} M(x, y) = c,$$

$$X = \text{those } x \text{ for which, for all } y \text{ in } Y,\ M(x, y) = c,$$

must be considered a routine computation in the present discussion, since any difficulty here will have arisen from the nature of the unspecified set A.

To complete the solution it is sufficient (a) to discover a finite subset $X' = \{x_i\}$ of X with

$$\min_{y \in B} \max_{x \in X'} M(x, y) = c$$

and then (b) to find weights λ_i for the x_i not more than $n - p$ of which are actually positive, and which make $\sum \lambda_i I_{x_i}$ an o.m.st. The process is best described geometrically.

Let y_0 designate a fixed interior point of Y, and let B' be some $(n - p - 1)$-dimensional cross section of B, meeting Y in precisely y_0. Each x of X describes a convex hyper-surface over B' which has one or more supporting hyper-planes at (y_0, c). Let S be a small sphere in B' with y_0 as center, and with each x associate the set S_x of points in S corresponding to the directions of steepest ascent of all the planes of support to $M(x, y)$ at (y_0, c). S_X will denote the union of the S_x for $x \in X$.[2]

The progress of the reduction may be traced through the following four statements which, for any fixed finite $X' \subseteq X$, are either all true or all false:

[2]If any plane is horizontal, the game is solved instantly, since the plane must correspond to a pure optimal x-strategy.

$$\min_{y \in B} \max_{x \in X'} M(x, y) = c \; ,$$

$$\min_{y \in B'} \max_{x \in X'} M(x, y) = c \; ,$$

(6)

$$\lim_{\rho \to 0} \min_{y \in S} \max_{x \in X'} M(x, y) = c, \quad \text{where } \rho \text{ is the radius of } S \; ,$$

$$\min_{y \in S} \max_{y' \in S_{X'}} y \cdot y' \geq 0 \; .$$

The inner product $y \cdot y'$ is taken relative to S as the unit sphere. Thus $y \cdot y'$ is the cosine of the angle $\measuredangle\, yy_0y'$.

Two assumptions must be interjected here:

(7) Y is not in the boundary of B ,

(8) $$\min_{y \in S} \max_{y' \in S_X} y \cdot y' = d > 0 \; .$$

Without the first, S would contain non-strategies. Without the other it becomes more difficult to show that the computation is finite. As will be seen in paragraph 7, failure of either (7) or (8) actually reduces the order of the optimal x-strategy, thus simplifying the computation. Geometrically, (8) states that y_0 is interior to the convex in B' spanned by S_X.

To continue: select y_1' at pleasure from S_X and proceed by the recursive instructions $(k = 2, 3, \ldots)$:

(I_k) If

$$m_k = \min_{y \in S} \max_{i < k} y \cdot y_i' < 0 \; ,$$

then let y_k denote the (unique) y in S at which the minimum occurs; if $m_k \geq 0$, terminate the process.

(II_k) Let y_k' denote a point of S_X for which[3]

$$y_k \cdot y_k' = \max_{y' \in S_X} y_k \cdot y' \; .$$

The iteration terminates after a finite number of steps. For if not, there would be $k_1 < k_2 < \ldots < k_l < \ldots$ for which the subsequences y_{k_l} and y_{k_l}' both converge. But then

$$\lim_{l \to \infty} m_{k_l} \geq \lim_{l \to \infty} y_{k_l} \cdot y_{k_{l-1}}' = \lim_{l \to \infty} y_{k_l} \cdot y_{k_l}' \geq d > 0$$

implies a finite termination after all.

---

[3]Note that S_X is a closed set..

The $\{y_1, \ldots, y_m\}$ so obtained leads back to a set $X' = \{x_1, \ldots, x_m\}$ for which the statements (6) are all true. Moreover a particular supporting plane $P_i(y)$ is denominated for each x_i. The supporting planes are distinct, but the x_i may not be.

The weights which solve the original game $M(x, y)$ will also solve the semi-discrete, linear game

$$P_i(y) = x_i \cdot y + c \quad (i = 1, \ldots, m;\ y \in S) ,$$

and conversely. The y-player here does not have a pure o.m.st. since the point y_0 is denied him, but any convex combination of $y \in S$ giving y_0 will be optimal, by the linearity. The linearity moreover makes it sufficient to consider the equivalent, wholly discrete game

$$||p_{ij}|| = ||P_i(y_j)|| \quad (i = 1, \ldots, m;\ j = 1, \ldots, n - p)$$

where the y_j are any $n - p$ points on the sphere S whose convex contains a neighborhood of the center, y_0. The y-player here has a unique o.m.st. with all weights positive. It follows (see [4]) that $m \geq n - p$ and that some $n - p \times n - p$ submatrix P' will have the property

$$\lambda_\nu = \frac{\sum_j P_{ji_\nu}}{\sum_\nu \sum_j P_{ji_\nu}} \geq 0, \quad \text{all } \nu = 1, 2, \ldots, n - p ,$$

P_{ji_ν} being the cofactor in P' of $P_{i_\nu j}$. P' may be discovered by a finite inspection. Then

$$F_0(x) = \sum_{\nu=1}^{n-p} \lambda_\nu I_{x_{i_\nu}}(x)$$

is the desired o.m.st.

§6. THE SOLUTION IN ONE DIMENSION

A complete description of the solution in the case $n = 2$ will serve to point up the discussion of the preceding section. For definiteness, let $M(x, y)$ be defined on the unit square $A \times B = [0, 1] \times [0, 1]$, and let it be continuous in each variable with the cross section at each x a convex curve over B.[4]

Suppose first that $\max_x M(x, y)$ has a unique minimum c at a point y_0 interior to the interval B. Then the set X of convex curves passing through (y_0, c) will be the union of two sets X_l and X_r, not necessarily disjoint, defined by:

---

[4] A. V. Martin collaborated with the authors in the original study of this case.

$$X_l = \text{those } x \in X \text{ with } M_l'(x, y_o) = \lim_{y \to y_o^-} \frac{M(x, y) - M(x, y_o)}{y - y_o} \leq 0 ,$$

$$X_r = \text{those } x \in X \text{ with } M_r'(x, y_o) = \lim_{y \to y_o^+} \frac{M(x, y) - M(x, y_o)}{y - y_o} \geq 0 .$$

To obtain an optimal x-strategy, select any $x_1 \in X_l$, $x_2 \in X_r$ and assign non-negative weights λ_1 and $\lambda_2 = 1 - \lambda_1$ satisfying

$$\lambda_1 M_l'(x_1, y_o) + \lambda_2 M_l'(x_2, y_o) \leq 0 \leq \lambda_1 M_r'(x_1, y_o) + \lambda_2 M_r'(x_2, y_o) .$$

These weights will be precisely determined only when $M(x_1, y)$ and $M(x_2, y)$ are actually differentiable at y_o. Otherwise there will be two extreme pairs of weights. Convex linear combinations of these extreme strategies, for all possible pairs $x_1, x_2 \in X_l, X_r$, will provide all o.m.st. of finite order for the x-player.

The same formulation is valid for $y_o = 0$ or $y_o = 1$ if the convention $M_l'(x, 0) = -\infty$, $M_r'(x, 1) = +\infty$ is adopted. In these cases $X_r \subseteq X_l$ and $X_l \subseteq X_r$ respectively; hence o.m.st. of order one, among others, will be found (cf. Corollary 1.2 above).

The same formulation is also valid trivially if $\max_x M(x, y)$ has its minimum over an interval Y. If y_o is any interior point of Y, then $M_l'(x, y_o) = M_r'(x, y_o) = 0$ for all $x \in X$. In this case all the extreme o.m.st. are pure.

EXAMPLE. Let $M(x, y) = f(y - x)$ in the unit square with $f''(u) > 0$ for $u \in [-1, 1]$. Suppose $f(-1) > f(0) < f(1)$, then the equation $f(u) = f(u - 1)$ has a unique solution $u = a$, $0 < a < 1$. In the light of the preceding discussion the following results may be stated:

(a) The value of the game is $f(a)$;

(b) The unique optimal y-strategy is I_a;

(c) The unique optimal x-strategy is $\alpha I_0 + (1 - \alpha) I_1$, where α is given by the equation $\alpha f'(a) + (1 - \alpha) f'(a - 1) = 0$.

If $f(-1) < f(0)$, or if $f(0) > f(1)$, then the unique optimal strategies are I_0 for both players, or I_1 for both players, respectively, and the value is $f(0)$. If $f(-1) = f(0)$ or $f(0) = f(1)$, or if one assumes only that $f''(u) \geq 0$, the optimal strategies are in general not unique.

§7. SHARPENING OF THE RESULTS

The discussion of this section will dispose of assumptions (7) and (8) of paragraph 5 and the proof of Corollary 1.2 of paragraph 3, and concurrently describe improved results for certain special situations.

First it may be remarked that, by using known properties of discrete games (see [4], and [1] Theorem 1), two sharper conclusions may be drawn from the matrix obtained in paragraph 5:

(i) every o.m.st. (of the discrete game) is a convex linear combination of extreme o.m.st. of order $n - p$ or less;

(ii) every strategy i participates in at least one such extreme o.m.st. of the x-player.

Referred to the original game, (i) implies that all o.m.st. of finite order may be put in terms of extreme o.m.st. of order $n - p$ or less. The construction of paragraph 5, of course, does not lead to a complete set of finite o.m.st. (to say nothing of the infinite ones that can easily be shown to exist whenever X is infinite). But, in consequence of (ii) and the arbitrariness of y_1' in paragraph 5, it will succeed in producing an extreme o.m.st. involving any one given x of X with positive weight.

Suppose now that Y is in the boundary of B, and hence that y_0 is in the boundary of B'. In order to contain the sphere S, B' must be enlarged. But if it is to become legal for the y-player to choose y from outside of B, it must also be made unprofitable, if the solution is not to be disrupted. Therefore, introduce a dummy strategy x_0 into the set A with pay-off

$$M(x_0, y) < c \quad \text{interior to } B ,$$

$$M(x_0, y) = c \quad \text{on boundary of } B ,$$

$$M(x_0, y) > c \quad \text{exterior to } B .$$

This function may be made continuous and convex in y since B is a convex region. It may also be made arbitrarily "steep" as it crosses the boundary, making it unimportant whether or not it is actually possible to extend the other functions $M(x, y)$ convexly into the exterior of B. Now by the remark of the last paragraph an o.m.st. of order $n - p$ or less may be found <u>utilizing</u> x_0 <u>with weight</u> $\lambda_0 > 0$. But the mixed strategy obtained by redistributing λ_0 among the other components, in proportion to their own weights, must be optimal in the original game.[5] Therefore, <u>at least one of the extreme o.m.st. is of order</u> $n - p - 1$ <u>or less</u>. Removed from the games context this conclusion becomes Corollary 1.2 of paragraph 3.

It might be remarked that a reduction of more than one — while possible — can not be deduced in general from the hypothesis that Y is situated in a lower-dimensional "corner" of the boundary of B.

To gather in the last loose end, suppose that assumption (8) of paragraph 5 does not hold. This would mean that along some directed line of B' emanating from y_0 none of the set of supporting planes actually increases. Equivalently, this would mean that the "bottom," Y, of the

[5]The formal proof is straightforward.

hyper-surface $z = \sup_x M(x, y)$ is less extensive than the "bottom," Y_L, of the envelope from above of the linear functions supporting $M(x, y)$ at Y. The prescription for dealing with this situation, should it occur, is simple: using Y_L in place of Y, define the cross section B'_L and sphere S_L. Then, replacing (8) with

$$\min_{y \in S_L} \max_{y' \in S_{LX}} y \cdot y' = d > 0 ,$$

proceed with the computation. The results involving $p = \dim Y$ will be replaced by stronger results involving $p_L = \dim Y_L < p$. Thus, unlike the boundary reduction detailed above, this case reduces <u>all</u> the extreme o.m.st. to order $n - p_L$ or less.

Finally, it is clear that the two reductions just described act independently, their effects being additive if both occur together.

BIBLIOGRAPHY

[1] BOHNENBLUST, H. F., KARLIN, S., SHAPLEY, L. S., "The Solutions of Discrete, Two-person Games," this Study.

[2] BLACKWELL, D., DRESHER, M., GIRSHICK, M. A., (private communication).

[3] MORGENSTERN, O., von NEUMANN, J., "Theory of Games and Economic Behavior," Ch. II, 2nd ed. (1947), Princeton University Press, Princeton.

[4] SHAPLEY, L. S., SNOW, R. N., "Basic Solutions of Discrete Games," this Study.

[5] VILLE, J., "Sur la Théorie Générale des Jeux où Intervient L'habilité des Joueurs," (in Traité du Calcul des Probabilitiés et de ses Applications by Emil Borel and collaborators, Vol. 2, No. 5) (1938), Paris.

[6] WALD, A., "Foundations of a General Theory of Sequential Decision Functions," Econometrica (1947), Vol. 15, pp. 279-313.

H. F. Bohnenblust

S. Karlin

L. S. Shapley

The RAND Corporation

BIBLIOGRAPHY

ALAOGLU, L. "Weak topologies of normed linear spaces," Annals of Math. (1940), 252-267.[1]

ALEXANDROFF, P. S., HOPF, H. Topologie, J. Springer, Berlin (1935), Anhang II.[1] (See Hopf, H., Alexandroff, P. S.)

ALLENDOERFER, C. B., WEIL, A. "The Gauss-Bonnet theorem for Riemannian polyhedra," Transactions of the American Mathematical Society 53 (1943), 104-112.[1] (See Weil, A., Allendoerfer, C. B.)

ANDERSON, O. "Theorie der Glücksspiele und ökonomisches Verhalten," Schweizerische Zeitschrift für Volkswirtschaft und Statistik 85 (1949), 46-53.[2]

ARROW, K. J., BLACKWELL, D., GIRSCHICK, M. A. "Bayes and Minimax solutions of sequential decision problems," Econometrica 17 (1949), 213-244. (See Blackwell, D., Girschick, M. A., Arrow, K. J. Also Girschick, M. A., Arrow, K. J., Blackwell, D.)

BELLMAN, R., BLACKWELL, D. "Some two-person games involving bluffing," Proceedings, National Academy of Sciences 35 (1949), 600-605. (See Blackwell, D., Bellman, R.)

BIRKHOFF, G. Lattice Theory, American Math. Soc. Colloq. Publ., 71-72.

BITTER, F. "The mathematical formulation of strategic problems," Proceedings of the Berkeley Symposium (ed. by J. Neyman), U. of Calif. Press (1949), 223-228.[2]

BLACKWELL, D., BELLMAN, R. "Some two-person games involving bluffing," Proceedings, National Academy of Sciences 35 (1949), 600-605. (See Bellman, R., Blackwell, D.)

BLACKWELL, D., GIRSCHICK, M. A., ARROW, K. J. "Bayes and Minimax solutions of sequential decision problems," Econometrica 17 (1949), 213-244. (See Arrow, K. J., Blackwell, D., Girschick, M. A. Also Girschick, M. A., Arrow, K. J., Blackwell, D.)

BOHNENBLUST, H. F., KARLIN, S., SHAPLEY, L. S. "Solutions of discrete two-person games," this Study. (See Karlin, S., Shapley, L. S., Bohnenblust, H. F. Also Shapley, L. S., Bohnenblust, H. F., Karlin, S.)

BOHNENBLUST, H. F., KARLIN, S. "On a theorem of Ville," this Study. (See Karlin, S., Bohnenblust, H. F.)

BOHNENBLUST, H. F., KARLIN, S., SHAPLEY, L. S. "Games with continuous, convex pay-off," this Study. (See Karlin, S., Shapley, L. S., Bohnenblust, H. F. Also Shapley, L. S., Bohnenblust, H. F., Karlin, S.)

BONESSEN, T., FENCHEL, W. Theorie der konvexen Körper, in Ergebnisse der Mathematik und ihre Grenzgebiete, Vol. III/1 (1934), Berlin.[1] (See Fenchel, W., Bonessen, T.)

BOREL, E. "Applications aux jeux de hasard," Traité du Calcul des Probabilités et de ses Applications, Gauthier-Villars, 4 f. II (1938), Paris.

BREMS, H. "Some notes on the structure of the duopoly problem," Nordisk Tidsskrift for Teknosk Økonomi, Nos. 1-4 (1948), 41-74.

BROWN, G. W., von NEUMANN, J. "Solutions of games by differential equations," this Study. (See von Neumann, J., Brown, G. W.)

BROWN, G. W. "Iterative solutions of games by ficitious play," Cowles Commission Monograph.[3]

BROWN, G. W., KOOPMANS, T. C. "Computational suggestions for maximizing a linear function subject to linear inequalities," Cowles Commission Monograph.[3] (See Koopmans, T. C., Brown, G. W.)

CHACKO, K. G. "Economic behaviour - A new theory," Indian Journal of Economics (in press).[2]

CHAMPERNOWNE, D. G. "A note on J. v. Neumann's article," Review of Economic Studies 13 (1) (1945-46), 10-18.

DANTZIG, G. B. "Application of the simplex method to a transportation problem," Cowles Commission Monograph.[3]

DANTZIG, G. B. "A proof of the equivalence of the programming problem and the game problem," Cowles Commission Monograph.[3]

DANTZIG, G. B. "Maximization of a linear function of variables subject to linear inequalities," Cowles Commission Monograph.[3]

DANTZIG, G. B. "Programming in a linear structure," Econometrica 17 (1949), 73-74.

DANTZIG, G. B., WOOD, M. K. "Programming of interdependent activities, I, General discussion," Econometrica 17 (1949), 193-199. (See Wood, M. K., Dantzig, G. B.)

DAVIES, D. W. "A theory of chess and noughts and crosses," Science News No. 16 (1950), 40-64.[2]

DEMARIA, G. "Su una nuova logica economica," Giornale Degli Economisti e Annali di Economia, VI (New series) (1947), 661-671.[2]

DINES, L. L. "Convex extension and linear inequalities," Bulletin of the American Mathematical Society 42 (1936), 353-365.[1]

DINES, L. L. "On a theorem of von Neumann," Proceedings, National Academy of Sciences, U.S.A., 33 (1947), 329-331.

DORFMAN, R. "Application of the simplex method to a game theory problem," Cowles Commission Monograph.[3]

DRESHER, M., KARLIN, S., SHAPLEY, L. S. "Polynomial games," this Study. (See Karlin, S., Shapley, L. S., Dresher, M. Also Shapley, L. S., Dresher, M., Karlin, S.)

DRESHER, M. "Methods of solution in game theory," Econometrica 18 (1950), 179-181.

DVORETZKY, A., WALD, A., WOLFOWITZ, J. "Elimination of randomization in certain problems of statistics and of the theory of games," Proceedings, National Academy of Sciences 36 (1950), 256-260. (See Wald, A., Wolfowitz, J., Dvoretzky, A. Also Wolfowitz, J., Dvoretzky, A., Wald, A.)

FARKAS, J. "Über die theorie der einfachen Ungleichungen," Journal für die reine und angewandte mathematik 124 (1901), 1-27.[1]

FENCHEL, W. "Krümmung und Windung Geschlossener Raumkurven," Math. Annalen 101 (1929), 238-252.[1]

FENCHEL, W., BONESSEN, T. Theorie der konvexen Körper, in Ergebnisse der Mathematik und ihre Grenzgebiete, Vol. III/1 (1934), Berlin.[1] (See Bonessen, T., Fenchel, W.)

FISHER, R. A. "Randomisation and an old enigma of card play," Mathematical Gazette 18 (1934), 294-297.

FRIEDMAN, M., SAVAGE, L. J. "The utility analysis of choices involving risk," Journal of Political Economy 56 (1948), 279-304.[2] (See Savage, L. J., Friedman, M.)

GALE, D., SHERMAN, S. "Solutions of finite two-person games," this Study. (See Sherman, S., Gale, D.)

GALE, D., KUHN, H. W., TUCKER, A. W. "On symmetric games," this Study. (See Kuhn, H. W., Tucker, A. W., Gale, D. Also Tucker, A. W., Gale, D., Kuhn, H. W.)

GALE, D., KUHN, H. W., TUCKER, A. W. "Reductions of game matrices," this Study. (See Kuhn, H. W., Tucker, A. W., Gale, D. Also Tucker, A. W., Gale, D., Kuhn, H. W.)

GALE, D. "Convex polyhedral cones and linear inequalities," Cowles Commission Monograph.[1, 3]

GALE, D., KUHN, H. W., TUCKER, A. W. "Linear programming and the theory of games," Cowles Commission Monograph.[3] (See Kuhn, H. W., Tucker, A. W., Gale, D. Also Tucker, A. W., Gale, D., Kuhn, H. W.)

GERSTENHABER, M. "Theory of convex polyhedral cones," Cowles Commission Monograph.[1, 3]

GIRSCHICK, M. A., ARROW, K. J., BLACKWELL, D. "Bayes and Minimax solutions of sequential decision problems," Econometrica 17 (1949), 213-244. (See Arrow, K. J., Blackwell, D., Girschick, M. A. Also Blackwell, D., Girschick, M. A., Arrow, K. J.)

GUILBAUD, G. T. "La théorie des jeux-contributions critiques à la théorie de la valeur," Economique Appliquée 2 (1949), 275-319.[2]

HARDY, G. H., LITTLEWOOD, J. E., POLYA, G. Inequalities, Cambridge Univ. Press (1934).[1] (See Littlewood, J. E., Polya, G., Hardy, G. H. Also Polya, G., Hardy, G. H., Littlewood, J. E.)

HITCHCOCK, F. L. "The distribution of a product from several sources to numerous localities," Journal of Mathematics and Physics 20 (1941), 224-230.

HOPF, H., ALEXANDROFF, P. S. Topologie, J. Springer, Berlin (1935), Anhang II.[1] (See Alexandroff, P. S., Hopf, H.)

HURWICZ, L. "The theory of economic behavior," American Economic Review 35 (1945), 909-925.[2]

JUSTMAN, E. "La théorie des jeux," (une nouvelle théorie de l'éliquilibre économique), Revue d'Economie Politique, nos. 5-6 (1949), 616-633.[2]

KAKUTANI, S. "A generalization of Brouwer's fixed point theorem," Duke Mathematical Journal 8 (1941), 457-459.

KAKUTANI, S. "Concrete representations of (M) spaces," Annals of Math. (1941), 994-1024.[1]

KALMAR, L. "Zur Theorie der abstrakten Spiele," Acta Szeged 4 (1928-29), 65-85.

KAPLANSKY, I. "A contribution to von Neumann's theory of games," Annals of Math. 46 (1945), 474-479.

KARLIN, S., SHAPLEY, L. S., BOHNENBLUST, H. F. "Solutions of discrete two-person games," this Study. (See Bohnenblust, H. F., Karlin, S., Shapley, L. S. Also Shapley, L. S., Bohnenblust, H. F., Karlin, S.)

KARLIN, S. "Operator treatment of minmax principle," this Study.

KARLIN, S., BOHNENBLUST, H. F. "On a theorem of Ville," this Study. (See Bohnenblust, H. F., Karlin, S.)

KARLIN, S., SHAPLEY, L. S., DRESHER, M. "Polynomial games," this Study. (See Dresher, M., Karlin, S., Shapley, L. S. Also Shapley, L. S., Dresher, M., Karlin, S.)

KARLIN, S., SHAPLEY, L. S., BOHNENBLUST, H. F. "Games with continuous, convex pay-off," this Study. (See Bohnenblust, H. F., Karlin, S., Shapley, L. S. Also Shapley, L. S., Bohnenblust, H. F., Karlin, S.)

KARLIN, S., SHAPLEY, L. S. "Geometry of reduced moment spaces," Proceedings, National Academy of Sciences 35 (1949), 673-679.[1] (See Shapley, L. S., Karlin, S.)

KAYSEN, C. "A revolution in economic theory?" The Review of Economic Studies 14/1 (1946-47), 1-15.[2]

KÖNIG, D. "Über eine Schlussweise aus dem Endlichen ins Unendliche," Acta Szeged 3 (1927), 121-130.

KOOPMANS, T. C., BROWN, G. W. "Computational suggestions for maximizing a linear function subject to linear inequalities," Cowles Commission Monograph.[3] (See Brown, G. W., Koopmans, T. C.)

KREIN, M., MILMAN, D. "On extreme points of regular convex sets," Studia Mathematica (1940), 133-138.[1] (See Milman, D., Krien, M.)

KUHN, H. W., TUCKER, A. W., GALE, D. "On symmetric games," this Study. (See Gale, D., Kuhn, H. W., Tucker, A. W. Also Tucker, A. W., Gale, D., Kuhn, H. W.)

KUHN, H. W., TUCKER, A. W., GALE, D. "Reductions of game matrices," this Study. (See Gale, D., Kuhn, H. W., Tucker, A. W. Also Tucker, A. W., Gale, D., Kuhn, H. W.)

KUHN, H. W. "A simplified two-person poker," this Study.

KUHN, H. W. "Extensive games," Proceedings, National Academy of Sciences 36 (1950).

KUHN, H. W., TUCKER, A. W., GALE, D. "Linear programming and the theory of games," Cowles Commission Monograph.[3] (See Gale, D., Kuhn, H. W., Tucker, A. W. Also Tucker, A. W., Gale, D., Kuhn, H. W.)

LEUNBACH, G. "Theory of games and economic behaviour," Nordisk Tidsskrift for Teknisk Økonomi, Nos. 1-4 (1948), 175-178.[2]

LITTLEWOOD, J. E., POLYA, G., HARDY, G. H. Inequalities, Cambridge Univ. Press (1934).[1] (See Hardy, G. H., Littlewood, J. E., Polya, G. Also Polya, G., Hardy, G. H., Littlewood, J. E.)

LOOMIS, L. H. "On a theorem of von Neumann," Proceedings, National Academy of Sciences 32 (1946), 213-215.

MARKOFF, A. "On mean values and exterior densities," Matematicheskii Sbornik (1938), 165-191.[1]

MARSCHAK, J. "Neumann's and Morgenstern's new approach to static economics," Journal of Political Economy 54 (1946), 97-115.[2]

MARSCHAK, J. "Rational behavior, uncertain prospects, and measurable utility," Econometrica 18 (1950), 111-141.[2]

McDONALD, J. "Poker: an American game," Fortune 37 (1948), 128-131 and 181-187.[2]

McDONALD, J. "The theory of strategy," Fortune 38 (1949), 100-110.[2]

McDONALD, J. Strategy in Poker, Business and War, W. W. Norton and Co., New York (1950).[2]

McKINSEY, J. C. C. "Isomorphism of games and strategic equivalence," this Study.

MENDEZ, J. "Progresos en la teoria economica de la conducta individual," Universidad Nacional de Columbia, Rivista Trimestial de Cultura Moderna 7 (1946), 259-276.[2]

MILMAN, D., KREIN, M. "On extreme points of regular convex sets," Studia Mathematica (1940), 133-138.[1] (See Krein, M., Milman, D.)

MINKOWSKI, H. Geometrie der Zahlen, Teubner, Leipzig (1910).[1]

MORGENSTERN, O., von NEUMANN, J. Theory of Games and Economic Behavior, Princeton (1944). (See von Neumann, J., Morgenstern, O.)

MORGENSTERN, O., von NEUMANN, J. Theory of Games and Economic Behavior, Princeton (1947, 2nd ed.). (See von Neumann, J., Morgenstern, O.)

MORGENSTERN, O. "Demand theory reconsidered," Quarterly Journal of Economics 62 (1948), 165-201.

MORGENSTERN, O. "Oligopoly, monopolistic competetion, and the theory of games," Proceedings, American Economic Review 38 (1948), 10-18.

MORGENSTERN, O. "Theorie des Spiels," Die Amerikanische Rundschau 5 (1949), 76-87.[2]

MORGENSTERN, O. "The theory of games," Scientific American 180 (1949), 22-25.[2]

MORGENSTERN, O. "Die Theorie der Spiele und des Wirtschaftlichen Verhaltens, Part I," Jahrbuch fur Sozialwissenschaft 1 (in press).[2]

MORGENSTERN, O. "Economics and the theory of games," Kyklos 4 (in press).[2]

MOTZKIN, T. S. Beiträge zur Theorie der linearen Ungleichungen, Inaug. diss., Jerusalem (1936).[1]

NASH, J. F., SHAPLEY, L. S., "A simple three-person poker game," this Study. (See Shapley, L. S., Nash, J. F.)

NASH, J. F. "Equilibrium points in n-person games," Proceedings, National Academy of Sciences 36 (1950), 48-49.

NASH, J. F. "The bargaining problem," Econometrica 18 (1950), 155-162.

von NEUMANN, J., BROWN, G. W. "Solutions of games by differential equations," this Study. (See Brown, G. W., von Neumann, J.)

von NEUMANN, J. "Zur Theorie der Gesellshaftsspiele," Mathematische Annalen 100 (1928), 295-320.

von NEUMANN, J. "Über ein ökonomisches Gleichungssystem und eine Verallgemeinerung des Brouwer'schen Fixpunktsatzes," Ergebnisse eines Math. Kolloquiums 8 (1937), 73-83.

von NEUMANN, J. "A model of general economic equilibrium," The Review of Economic Studies 13 (1945-46), 1-9.

von NEUMANN, J., MORGENSTERN, O. Theory of Games and Economic Behavior, Princeton (1944). (See Morgenstern, O., von Neumann, J.)

von NEUMANN, J., MORGENSTERN, O. Theory of Games and Economic Behavior, Princeton (1947, 2nd ed.). (See Morgenstern, O., von Neumann, J.)

PAXSON, E. W. "Recent developments in the mathematical theory of games," Econometrica 17 (1949), 72-73.[2]

POLYA, G., HARDY, G. H., LITTLEWOOD, J. E. Inequalities, Cambridge Univ. Press (1934).[1] (See Hardy, G. H., Littlewood, J. E., Polya, G. Also Littlewood, J. E., Polya, G., Hardy, G. H.)

POPOVICIU, T. "Les fonctions convexes," Actualités Scientifiques, no. 992, Hermann, Paris (1944).[1]

de POSSEL, R. "Sur la théorie mathématique des jeux de hasard et de reflexion," Actualités Scientifiques et Industrielles, no. 436, Hermann, Paris (1936).[2]

RICHARDSON, M. "On weakly ordered systems," Bulletin of American Mathematical Society 52 (1946), 113-116.

RUIST, E. "Speltieri och ekonomiska problem," Economisk Tidskrift 2 (1949), 112-117.[2]

SAVAGE, L. J., FRIEDMAN, M. "The utility analysis of choices involving risk," Journal of Political Economy 56 (1948), 279-304.[2] (See Friedman, M., Savage, L. J.)

SCHAUDER, J. "Der Fixpunktsatz in Funktionalraümen," Studia Mathematica, Vol. 2 (1930), 171-180.[1]

SHAPLEY, L. S., SNOW, R. N. "Basic solutions of discrete games," this Study. (See Snow, R. N., Shapley, L. S.)

SHAPLEY, L. S., BOHNENBLUST, H. F., KARLIN, S. "Solutions of discrete two-person games," this Study. (See Bohnenblust, H. F., Karlin, S., Shapley, L. S. Also Karlin, S., Shapley, L. S., Bohnenblust, H. F.)

SHAPLEY, L. S., NASH, J. F. "A simple three-person poker game," this Study. (See Nash, J. F., Shapley, L. S.)

SHAPLEY, L. S., DRESHER, M., KARLIN, S. "Polynomial games," this Study. (See Dresher, M., Karlin, S., Shapley, L. S. Also Karlin, S., Shapley, L. S., Dresher, M.)

SHAPLEY, L. S., BOHNENBLUST, H. F., KARLIN, S. "Games with continuous, convex pay-off," this Study. (See Bohnenblust, H. F., Karlin, S., Shapley, L. S. Also Karlin, S., Shapley, L. S., Bohnenblust, H. F.)

SHAPLEY, L. S., KARLIN, S. "Geometry of reduced moment spaces," Proceedings, National Academy of Sciences 35 (1949), 673-679.[1] (See Karlin, S., Shapley, L. S.)

SHERMAN, S., GALE, D. "Solutions of finite two-person games," this Study. (See Gale, D., Sherman, S.)

SHERMAN, S. "Games and sub-games," Proceedings, American Math. Soc. (1951).

SHOHAT, J. A., TAMARKIN, J. D. The Problem of Moments, Amer. Math. Soc. (1943), New York.[1] (See Tamarkin, J. D., Shohat, J. A.)

SINGER, K. "Robot economics," Economic Record 25 (1949), 48-73.

SNOW, R. N., SHAPLEY, L. S. "Basic solutions of discrete games," this Study. (See Shapley, L. S., Snow, R. N.)

STIEMKE, E. "Uber positive Losungen homogener linearer Gleichungen," Mathematische Annalen 76 (1915), 340-342.[1]

STONE, R. "The theory of games," Economic Journal 58 (1948), 185-201.[2]

Symposium of Statistical Inference in Decision Making (Dvoretzky, Wald, Chernoff, Savage, Sobel), Econometrica 18 (1950), 181-185.

TAMARKIN, J. D., SHOHAT, J. A. The Problem of Moments, Amer. Math. Soc. (1943), New York.[1] (See Shohat, J. A., Tamarkin, J. D.)

TUCKER, A. W., GALE, D., KUHN, H. W. "On symmetric games," this Study. (See Gale, D., Kuhn, H. W., Tucker, A. W. Also Kuhn, H. W., Tucker, A. W., Gale, D.)

TUCKER, A. W., GALE, D., KUHN, H. W. "Reductions of game matrices," this Study. (See Gale, D., Kuhn, H. W., Tucker, A. W. Also Kuhn, H. W., Tucker, A. W., Gale, D.)

TUCKER, A. W., GALE, D., KUHN, H. W. "Linear programming and the theory of games," Cowles Commission Monograph.[3] (See Gale, D., Kuhn, H. W., Tucker, A. W. Also Kuhn, H. W., Tucker, A. W., Gale, D.)

TUKEY, J. W. "A problem in strategy," Econometrica 17 (1949), 73.

VILLE, J. "Sur la théorie générale des jeux où intervient l'habilité des joueurs," Traité du Calcul des Probabilités et de ses Applications, by E. Borel and collaborators, Paris (1938), Vol. 2, No. 5, 105-113.

WALD, A. "Ueber die eindeutige positive Loesbarkeit der neuen Productionsgleichungen," Ergebnisse eines Mathematischen Kolloquiums 6 (1935), with discussion by K. Menger, 12-20.

WALD, A. "Ueber die Produktionsgleichungen der Oekonomischen Wertlehre (II. Mitteilung)," Ergebnisse eines Mathematischen Kolloquiums 7 (1936), 1-6.

WALD, A. "Ueber Einige Gleichungssysteme der Mathematischen Oekonomie," Zeitschrift Fur Nationaloekonomie 7 (1936), 637-670.

WALD, A. "Generalization of a theorem by von Neumann concerning zero sum two-person games," Annals of Mathematics 46 (1945), 281-286.

WALD, A. "Statistical decision functions which minimize the maximum risk," Annals of Mathematics 46 (1945), 265-280.

WALD, A. "Foundation of a general theory of sequential decision functions," Econometrica 15 (1947), 279-313.

WALD, A. "Statistical decision functions," Annals of Mathematical Statistics 20 (1949), 165-205.

WALD, A., WOLFOWITZ, J., DVORETZKY, A. "Elimination of randomization in certain problems of statistics and of the theory of games," Proceedings, National Academy of Sciences 36 (1950), 256-260. (See Dvoretzky, A., Wald, A., Wolfowitz, J. Also Wolfowitz, J., Dvoretzky, A., Wald, A.)

WALD, A. Statistical Decision Functions, John Wiley and Sons, N. Y. (1950).

WEIL, A., ALLENDOERFER, C. B. "The Gauss-Bonnet theorem for Riemannian polyhedra," Transactions of the American Mathematical Society 53 (1943), 104-112.[1] (See Allendoerfer, C. B., Weil, A.)

WEINBERGER, O. "Wirtschaftshandlungen und Spielstrategie," Statistische Vierteljahresschrift 2 (1949), 24-31.[2]

WEYL, H. "The elementary theory of convex polyhedra," this Study.

WEYL, H. "Elementary proof of a minimax theorem due to von Neumann," this Study.

WOLFOWITZ, J., DVORETZKY, A., WALD, A. "Elimination of randomization in certain problems of statistics and of the theory of games," Proceedings, National Academy of Sciences 36 (1950), 256-260. (See Dvoretzky, A., Wald, A., Wolfowitz, J. Also Wald, A., Wolfowitz, J., Dvoretzky, A.)

WOOD, M. K., DANTZIG, G. B. "Programming of interdependent activities, I, General discussion," Econometrica 17 (1949), 193-199. (See Dantzig, G. B., Wood, M. K.)

NOTE

A supplementary Bibliography will be found in Annals of Mathematics Study No. 28, CONTRIBUTIONS TO THE THEORY OF GAMES, Volume II. This includes more recent references, as well as some omissions which were drawn to the attention of the Editors.

[1]Source papers, primarily from the theory of linear inequalities.

[2]Book reviews and expository articles.

[3]"Activity Analysis of Production and Allocation," Cowles Commission Monograph No. 13, John Wiley and Sons, New York (in preparation).

Ingram Content Group UK Ltd.
Milton Keynes UK
UKHW032005260523
422428UK00001B/10

9 780691 079349